Communications
in Computer and Information Science   2031

Rationale

The CCIS series is devoted to the publication of proceedings of computer science conferences. Its aim is to efficiently disseminate original research results in informatics in printed and electronic form. While the focus is on publication of peer-reviewed full papers presenting mature work, inclusion of reviewed short papers reporting on work in progress is welcome, too. Besides globally relevant meetings with internationally representative program committees guaranteeing a strict peer-reviewing and paper selection process, conferences run by societies or of high regional or national relevance are also considered for publication.

Topics

The topical scope of CCIS spans the entire spectrum of informatics ranging from foundational topics in the theory of computing to information and communications science and technology and a broad variety of interdisciplinary application fields.

Information for Volume Editors and Authors

Publication in CCIS is free of charge. No royalties are paid, however, we offer registered conference participants temporary free access to the online version of the conference proceedings on SpringerLink (http://link.springer.com) by means of an http referrer from the conference website and/or a number of complimentary printed copies, as specified in the official acceptance email of the event.

CCIS proceedings can be published in time for distribution at conferences or as post-proceedings, and delivered in the form of printed books and/or electronically as USBs and/or e-content licenses for accessing proceedings at SpringerLink. Furthermore, CCIS proceedings are included in the CCIS electronic book series hosted in the SpringerLink digital library at http://link.springer.com/bookseries/7899. Conferences publishing in CCIS are allowed to use Online Conference Service (OCS) for managing the whole proceedings lifecycle (from submission and reviewing to preparing for publication) free of charge.

Publication process

The language of publication is exclusively English. Authors publishing in CCIS have to sign the Springer CCIS copyright transfer form, however, they are free to use their material published in CCIS for substantially changed, more elaborate subsequent publications elsewhere. For the preparation of the camera-ready papers/files, authors have to strictly adhere to the Springer CCIS Authors' Instructions and are strongly encouraged to use the CCIS LaTeX style files or templates.

Abstracting/Indexing

CCIS is abstracted/indexed in DBLP, Google Scholar, EI-Compendex, Mathematical Reviews, SCImago, Scopus. CCIS volumes are also submitted for the inclusion in ISI Proceedings.

How to start

To start the evaluation of your proposal for inclusion in the CCIS series, please send an e-mail to ccis@springer.com.

Kanubhai K. Patel · KC Santosh · Atul Patel ·
Ashish Ghosh
Editors

Soft Computing and Its Engineering Applications

5th International Conference, icSoftComp 2023
Changa, Anand, India, December 7–9, 2023
Revised Selected Papers, Part II

 Springer

Editors
Kanubhai K. Patel ⓘ
Charotar University of Science
and Technology
Changa, India

Atul Patel ⓘ
Charotar University of Science
and Technology
Changa, India

KC Santosh ⓘ
University of South Dakota
Vermillion, SD, USA

Ashish Ghosh ⓘ
Indian Statistical Institute
Kolkata, India

ISSN 1865-0929   ISSN 1865-0937 (electronic)
Communications in Computer and Information Science
ISBN 978-3-031-53727-1   ISBN 978-3-031-53728-8 (eBook)
https://doi.org/10.1007/978-3-031-53728-8

This Springer imprint is published by the registered company Springer Nature Switzerland AG
The registered company address is: Gewerbestrasse 11, 6330 Cham, Switzerland

Paper in this product is recyclable.

Preface

It is a matter of great privilege to have been tasked with the writing of this preface for the proceedings of The Fifth International Conference on Soft Computing and its Engineering Applications (icSoftComp 2023). The conference aimed to provide an excellent international forum for emerging and accomplished research scholars, academicians, students, and professionals in the areas of computer science and engineering to present their research, knowledge, new ideas, and innovations. The conference was held during 07–09 December 2023, at Charotar University of Science & Technology (CHARUSAT), Changa, India, and organized by the Faculty of Computer Science and Applications, CHARUSAT. There are three pillars of Soft Computing viz., i) Fuzzy computing, ii) Neuro computing, and iii) Evolutionary computing. Research submissions in these three areas were received. The Program Committee of icSoftComp 2023 is extremely grateful to the authors from 17 different countries including UK, USA, France, Germany, Portugal, North Macedonia, Tunisia, Lithuania, United Arab Emirates, Sharjah, Saudi Arabia, Bangladesh, Philippines, Finland, Malaysia, South Africa, and India who showed an overwhelming response to the call for papers, submitting over 351 papers. The entire review team (Technical Program Committee members along with 14 additional reviewers) expended tremendous effort to ensure fairness and consistency during the selection process, resulting in the best-quality papers being selected for presentation and publication. It was ensured that every paper received at least three, and in most cases four, reviews. Checking of similarities was also done based on international norms and standards. After a rigorous peer review 44 papers were accepted with an acceptance ratio of 12.54%. The papers are organised according to the following topics: Theory & Methods, Systems & Applications, and Hybrid Techniques. The proceedings of the conference are published as two volumes in the Communications in Computer and Information Science (CCIS) series by Springer, and are also indexed by ISI Proceedings, DBLP, Ulrich's, EI-Compendex, SCOPUS, Zentralblatt Math, MetaPress, and SpringerLink. We, in our capacity as volume editors, convey our sincere gratitude to Springer for providing the opportunity to publish the proceedings of icSoftComp 2023 in their CCIS series.

icSoftComp 2023 exhibited an exciting technical program. It also featured high-quality workshops, two keynotes and six expert talks from prominent research and industry leaders. Keynote speeches were given by Theofanis P. Raptis (CNR, Italy), Sardar Islam (Victoria University, Australia), and Massimiliano Cannata (SUPSI, Switzerland). Expert talks were given by Xun Shao (Toyohashi University of Technology, Japan), Ashis Jalote Parmar (NTNU, Norway), Chang Yoong Choon (Universiti Tunku Abdul Rahman, Malaysia), Biplab Banerjee (IIT Bombay, India), Sonal Jain (SPU, India), and Sharnil Pandya (Linnaeus University, Sweden). We are grateful to them for sharing their insights on their latest research with us.

The Organizing Committee of icSoftComp 2023 is indebted to R.V. Upadhyay, Provost of Charotar University of Science and Technology and Patron, for the confidence that he invested in us in organizing this international conference. We would also like to take this opportunity to extend our heartfelt thanks to the honorary chairs of this conference, Kalyanmoy Deb (Michigan State University, MI, USA), Witold Pedrycz (University of Alberta, Alberta, Canada), Leszek Rutkowski (IEEE Fellow) (Czesto-chowa University of Technology, Czestochowa, Poland), and Janusz Kacprzyk (Polish Academy of Sciences, Warsaw, Poland) for their active involvement from the very begin-ning until the end of the conference. The quality of a refereed volume primarily depends on the expertise and dedication of the reviewers, who volunteer with a smiling face. The editors are further indebted to the Technical Program Committee members and exter-nal reviewers who not only produced excellent reviews but also did so in a short time frame, in spite of their very busy schedules. Because of their quality work it was possible to maintain the high academic standard of the proceedings. Without their support, this conference could never have assumed such a successful shape. Special words of appre-ciation are due to note the enthusiasm of all the faculty, staff, and students of the Faculty of Computer Science and Applications of CHARUSAT, who organized the conference in a professional manner.

It is needless to mention the role of the contributors. The editors would like to take this opportunity to thank the authors of all submitted papers not only for their hard work but also for considering the conference a viable platform to showcase some of their latest findings, not to mention their adherence to the deadlines and patience with the tedious review process. Special thanks to the team of EquinOCS, whose paper submission plat-form was used to organize reviews and collate the files for these proceedings. We also wish to express our thanks to Amin Mobasheri (Editor, Computer Science Proceedings, Springer Heidelberg) for his help and cooperation. We gratefully acknowledge the finan-cial (partial) support received from Department of Science & Technology, Government of India and Gujarat Council on Science & Technology (GUJCOST), Government of Gujarat, Gandhinagar, India for organizing the conference. Last but not least, the editors profusely thank all who directly or indirectly helped us in making icSoftComp 2023 a grand success and allowed the conference to achieve its goals, academic or otherwise.

December 2023

Kanubhai K. Patel
KC Santosh
Atul Patel
Ashish Ghosh

Organization

Patron

R. V. Upadhyay   Charotar University of Science and Technology, India

Honorary Chairs

Kalyanmoy Deb   Michigan State University, USA
Witold Pedrycz   University of Alberta, Canada
Leszek Rutkowski   Czestochowa University of Technology, Poland
Janusz Kacprzyk   Polish Academy of Sciences, Poland

General Chairs

Atul Patel   Charotar University of Science and Technology, India

George Ghinea   Brunel University London, UK
Dilip Kumar Pratihar   Indian Institute of Technology Kharagpur, India
Pawan Lingras   Saint Mary's University, Canada

Technical Program Committee Chair

Kanubhai K. Patel   Charotar University of Science and Technology, India

Technical Program Committee Co-chairs

Ashish Ghosh   Indian Statistical Institute, Kolkata, India
KC Santosh   University of South Dakota, USA
Deepak Garg   Bennett University, India
Gayatri Doctor   CEPT University, India

| Maryam Kaveshgar | Ahmedabad University, India |
| Ashis Jalote-Parmar | Norwegian University of Science and Technology, Norway |

Advisory Committee

Arup Dasgupta	Geospatial Media and Communications, India
Valentina E. Balas	University of Arad, Romania
Bhuvan Unhelkar	University of South Florida Sarasota-Manatee, USA
Dharmendra T. Patel	Charotar University of Science and Technology, India
Indrakshi Ray	Colorado State University, USA
J. C. Bansal	Soft Computing Research Society, India
Narendra S. Chaudhari	Indian Institute of Technology Indore, India
Rajendra Akerkar	Vestlandsforsking, Norway
Sudhir Kumar Barai	BITS Pilani, India
S. P. Kosta	Charotar University of Science and Technology, India

Technical Program Committee Members

Abhijit Datta Banik	IIT Bhubaneswar, India
Abdulla Omeer	Dr. Babasaheb Ambedkar Marathwada University, India
Abhineet Anand	Chitkara University, India
Aditya Patel	Kamdhenu University, India
Adrijan Božinovski	University American College Skopje, North Macedonia
Aji S.	University of Kerala, India
Akhil Meerja	Vardhaman College of Engineering, India
Aman Sharma	Jaypee University of Information Technology, India
Ami Choksi	C.K. Pithawala College of Engg. and Technology, India
Amit Joshi	Malaviya National Institute of Technology, India
Amit Thakkar	Charotar University of Science and Technology, India
Amol Vibhute	Symbiosis Institute of Computer Studies and Research, India
Anand Nayyar	Duy Tan University, Vietnam

Angshuman Jana	IIIT Guwahati, India
Ansuman Bhattacharya	IIT (ISM) Dhanbad, India
Anurag Singh	IIIT-Naya Raipur, India
Aravind Rajam	Washington State University, Pullman, USA
Arjun Mane	Government Institute of Forensic Science, India
Arpankumar Raval	Charotar University of Science and Technology, India
Arti Jain	Jaypee Institute of Information Technology, India
Arunima Jaiswal	Indira Gandhi Delhi Technical University for Women, India
Asha Manek	RVITM Engineering College, India
Ashok Patel	Florida Polytechnic University, USA
Ashok Sharma	Lovely Professional University, India
Ashraf Elnagar	University of Sharjah, UAE
Ashutosh Kumar Dubey	Chitkara University, India
Ashwin Makwana	Charotar University of Science and Technology, India
Avimanyou Vatsa	Fairleigh Dickinson University - Teaneck, USA
Avinash Kadam	Dr. Babasaheb Ambedkar Marathwada University, India
Ayad Mousa	University of Kerbala, Iraq
Bhaskar Karn	BIT Mesra, India
Bhavik Pandya	Navgujarat College of Computer Applications, India
Bhogeswar Borah	Tezpur University, India
Bhuvaneswari Amma	IIIT Una, India
Chaman Sabharwal	Missouri University of Science and Technology, USA
Charu Gandhi	Jaypee University of Information Technology, India
Chirag Patel	Innovate Tax, UK
Chirag Paunwala	SCET, India
Costas Vassilakis	University of the Peloponnese, Greece
Darshana Patel	Rai University, India
Dattatraya Kodavade	DKTE Society's Textile and Engineering Institute, India
Dayashankar Singh	Madan Mohan Malaviya University of Technology, India
Deepa Thilak	SRM University, India
Deepak N. A.	RV Institute of Technology and Management, India
Deepak Singh	IIIT, Lucknow, India
Delampady Narasimha	IIT Dharwad, India

Dharmendra Bhatti	Uka Tarsadia University, India
Digvijaysinh Rathod	National Forensic Sciences University, India
Dinesh Acharya	Manipal Institute of Technology, India
Divyansh Thakur	IIIT Una, India
Dushyantsinh Rathod	Alpha College of Engineering and Technology, India
E. Rajesh	Galgotias University, India
Gururaj Mukarambi	Central University of Karnataka, India
Gururaj H. L.	Vidyavardhaka College of Engineering, India
Hardik Joshi	Gujarat University, India
Harshal Arolkar	GLS University, India
Himanshu Jindal	Jaypee University of Information Technology, India
Hiren Joshi	Gujarat University, India
Hiren Mewada	Prince Mohammad Bin Fahd University, Saudi Arabia
Irene Govender	University of KwaZulu-Natal, South Africa
Jagadeesha Bhatt	IIIT Dharwad, India
Jaimin Undavia	Charotar University of Science and Technology, India
Jaishree Tailor	Uka Tarsadia University, India
Janmenjoy Nayak	AITAM, India
Jaspher Kathrine	Karunya Institute of Technology and Sciences, India
Jimitkumar Patel	Charotar University of Science and Technology, India
Joydip Dhar	ABV-IIITM, India
József Dombi	University of Szeged, Hungary
Kamlendu Pandey	VNSGU, India
Kamlesh Dutta	NIT Hamirpur, India
Kiran Trivedi	Northeastern University, USA
KiranSree Pokkuluri	Shri Vishnu Engineering College for Women, India
Krishan Kumar	National Institute of Technology Uttarakhand, India
Kuldip Singh Patel	IIIT Naya Raipur, India
Kuntal Patel	Ahmedabad University, India
Latika Singh	Sushant University, India
M. Srinivas	National Institute of Technology-Warangal, India
M. A. Jabbar	Vardhaman College of Engineering, India
Maciej Ławrynczuk	Warsaw University of Technology, Poland
Mahmoud Elish	Gulf University for Science and Technology, Kuwait

Mandeep Kaur	Sharda University, India
Manoj Majumder	IIIT Naya Raipur, India
Meera Kansara	Gujarat Vidyapith, India
Michał Chlebiej	Nicolaus Copernicus University, Poland
Mittal Desai	Charotar University of Science and Technology, India
Mohamad Ijab	National University of Malaysia, Malaysia
Mohini Agarwal	Amity University Noida, India
Monika Patel	NVP College of Pure and Applied Sciences, India
Mukti Jadhav	Marathwada Institute of Technology, India
Neetu Sardana	Jaypee University of Information Technology, India
Nidhi Arora	Solusoft Technologies Pvt. Ltd., India
Nilay Vaidya	Charotar University of Science and Technology, India
Nitin Kumar	National Institute of Technology Uttarakhand, India
Parag Rughani	GFSU, India
Parul Patel	VNSGU, India
Prashant Pittalia	Sardar Patel University, India
Priti Sajja	Sardar Patel University, India
Pritpal Singh	Jagiellonian University, Poland
Punya Paltani	IIIT Naya Raipur, India
Rajeev Kumar	NIT Hamirpur, India
Rajesh Thakker	Vishwakarma Govt Engg College, India
Ramesh Prajapati	LJ Institute of Engineering and Technology, India
Ramzi Guetari	University of Tunis El Manar, Tunisia
Rana Mukherji	ICFAI University, Jaipur, India
Rashmi Saini	GB Pant Institute of Engineering and Technology, India
Rathinaraja Jeyaraj	National Institute of Technology Karnataka, India
Rekha A. G.	State Bank of India, India
Rohini Rao	Manipal Academy of Higher Education, India
S. Shanmugam	Concordia University Chicago, USA
S. Srinivasulu Raju	VR Siddhartha Engineering College, India
Sailesh Iyer	Rai University, India
Saman Chaeikar	Iranians University e-Institute of Higher Education, Iran
Sameerchand Pudaruth	University of Mauritius, Mauritius
Samir Patel	PDPU, India
Sandeep Gaikwad	Symbiosis Institute of Computer Studies and Research, India
Sandhya Dubey	Manipal Academy of Higher Education, India

Sanjay Moulik	IIIT Guwahati, India
Sannidhan M. S.	NMAM Institute of Technology, India
Sanskruti Patel	Charotar University of Science and Technology, India
Saurabh Das	University of Calcutta, India
S. B. Goyal	City University of Malaysia, Malaysia
Shachi Sharma	South Asian University, India
Shailesh Khant	Charotar University of Science and Technology, India
Shefali Naik	Ahmedabad University, India
Shilpa Gite	Symbiosis Institute of Technology, India
Shravan Kumar Garg	Swami Vivekanand Subharti University, India
Sohil Pandya	Charotar University of Science and Technology, India
Spiros Skiadopoulos	University of the Peloponnese, Greece
Srinibas Swain	IIIT Guwahati, India
Srinivasan Sriramulu	Galgotias University, India
Subhasish Dhal	IIIT Guwahati, India
Sudhanshu Maurya	Graphic Era Hill University, Malaysia
Sujit Das	National Institute of Technology-Warangal, India
Sumegh Tharewal	Dr. Babasaheb Ambedkar Marathwada University, India
Sunil Bajeja	Marwadi University, India
Swati Gupta	Jaypee University of Information Technology, India
Tanima Dutta	Indian Institute of Technology (BHU), India
Tanuja S. Dhope	Rajarshi Shahu College of Engineering, India
Thoudam Singh	NIT Silchar, India
Tzung-Pei Hong	National University of Kaohsiung, Taiwan
Vana Kalogeraki	Athens University of Economics and Business, Greece
Vasudha M. P.	Jain University, India
Vatsal Shah	BVM Engineering, India
Veena Jokhakar	VNSGU, India
Vibhakar Pathak	Arya College of Engg. and IT, India
Vijaya Rajanala	SR Engineering College, India
Vinay Vachharajani	Ahmedabad University, India
Vinod Kumar	IIIT Lucknow, India
Vishnu Pendyala	San José State University, USA
Yogesh Rode	Jijamata Mahavidhyalaya Buldana, India
Zina Miled	Indiana University, USA

Additional Reviewers

Anjali Mahavar
Falguni Parsana
Himanshu Patel
Kamalesh Salunke
Krishna Kant
Lokesh Sharma
Mihir Mehta

Parag Shukla
Prashant Dolia
Shanti Verma
Tejasvi Koti
Rachana Parikh
Ramesh Chandra Goswami
Rayeesa Tasneem

Contents – Part II

Hybrid Techniques

Contents – Part I

System and Applications

Systems and Applications

Metagenomic Gene Prediction Using Bidirectional LSTM

K. Syama[✉] and J. Angel Arul Jothi

Department of Computer Science, Birla Institute of Technology and Science
Pilani Dubai Campus, Dubai, UAE
{p20190011,angeljothi}@dubai.bits-pilani.ac.in

Abstract. Genomics and microbial research have been revolutionized by the beginning of low-cost, high-throughput sequencing technologies. Which, in turn, brings a large amount of genomes to public archives today. Annotation tools are essential to understanding these microorganisms. The metagenomic sequences are fragmented, which makes accurate gene prediction challenging. Most computational gene predictor models use machine learning (ML) and deep learning (DL) to predict genes in metagenomic sequences. However, to capture the sequential dependencies and contextual information within the sequences, recurrent neural networks are more popular. This study uses a bi-directional long short-term memory (LSTM) model to classify input ORF sequences into coding or non-coding classes. The proposed model is compared with other DL methods, such as convolutional neural networks (CNN) and LSTM models. It achieved an area under the curve (AUC) value of 99%, Accuracy of 95.3%, Precision of 96.53%, Recall of 94.57% and F1-score of 95.22%.

Keywords: Gene prediction · long short-term memory · Metagenomics · Bi-directional LSTM

1 Introduction

Metagenomics is a genomics field involving the study of genetic material collected from environmental samples [1]. Metagenomics has many applications in forensics, engineering, ecology, discovering new antibiotics, personalized medicine, etc. Gene prediction identifies the parts of deoxyribonucleic acid (DNA) sequence that can encode a gene. It is also known as gene annotation or gene finding. Traditionally, identification uses wet lab experiments, which involve experiments on living cells and organisms. The wet lab experiments often involved labor-intensive and expensive processes in terms of both time and money. Hence, computational approaches are required to predict genes. There are two types of computational metagenomic gene prediction approaches: content-based and similarity-based [2]. In similarity-based approaches, the similarity between the existing and candidate genes is calculated. The content-based approaches extract different features and apply supervised learning techniques. Hence, content-based approaches take less time for the computation.

© The Author(s), under exclusive license to Springer Nature Switzerland AG 2024
K. K. Patel et al. (Eds.): icSoftComp 2023, CCIS 2031, pp. 3–15, 2024.
https://doi.org/10.1007/978-3-031-53728-8_1

Metagenomic gene prediction is challenging as the sequences are incomplete, short and fragmented [3]. Applying machine learning (ML) algorithms for gene prediction produce promising results, though the sequences are fragmented. ML-based gene predictor models try to learn the characteristics of open reading frames (ORFs) by extracting features from the ORFs [4]. An ORF is a sequence of DNA or ribonucleic acid (RNA) that has the potential to be translated into a functional protein. Each ORF sequence can be classified as coding or non-coding. A coding ORF has the potential to encode a functional protein, whereas a non-coding ORF does not encode a functional protein. The extracted features included monocodon usage, dicodon usage, translation initiation sites (TISs), ORF length, and guanine or cytosine (GC) content.

DL techniques have recently been used to investigate metagenomic raw sequence data. Various DL methods like CNNs and deep neural networks (DNNs) have achieved considerable attention in gene prediction problems [5,6]. In addition, recurrent neural networks (RNNs) have proven their success in learning and processing sequence data. Their architecture is designed to capture the temporal dependencies in the sequence. RNNs include variations like LSTM networks, gated recurrent units (GRUs), bidirectional LSTM (Bi-LSTM), bidirectional GRU and Transformers.

LSTMs are a robust variant of RNNs and are particularly well-suited for tasks involving sequences by avoiding the vanishing and exploding gradient problems of traditional RNNs by introducing a more complex memory cell structure. Bi-LSTMs are extension of LSTMs, and it is able to capture the sequential information by simultaneously processing the input sequence in forward and backward directions.

This work puts forward the following contributions.

1. A gene predictor model is proposed using Bi-LSTM to classify the input ORF sequences to coding or non-coding.
2. The various hyperparameters of the Bi-LSTM model, such as the number of hidden layers and the number of units in each layer, optimizer, and dropout rate, are tuned using a manual grid search with a predefined grid of parameters to decide the best values for the parameters. Early stopping and model checkpoints are used to get the best model during hyperparameter tuning.
3. The outcomes of the Bi-LSTM model proposed in this study are compared against an existing model and other RNN models like LSTM and CNN.

The paper is organized as follows. Section 2 briefly explains previous works done in the gene prediction domain. Section 3 explains the detailed methodology of this work. Followed by the methodology, the setup done for the experimentation, and the metrics used to evaluate the work are explained. After this, the findings are explored. Finally, the conclusion and the future directions are explained.

2   Related Works

Numerous studies have been conducted using ML methods for effective and efficient gene prediction. This review analyzes the various studies that classified the ORF sequences as coding or non-coding. In [7], a two-stage ML approach was proposed for gene prediction. This work used linear discriminant for feature extraction and neural networks to find the probability of encoding a protein. The model achieved 89% sensitivity, 93% specificity and an average accuracy of 98%.

In [8], a set of related features was selected from the entire set of features using minimum redundancy maximum relevance(mRMR). SVM was trained, and it gave the probabilities of the coding class. The model was compared with mRMR-neural network model, orphelia, MGC and Prodigal. The model had achieved 92.17% harmonic mean, specificity of 96.53%, and sensitivity of 88.31%. The proposed technique outperformed Orphelia and MGC by an average of 11%.

A novel method called GeneRFinder was proposed by [9], which used an RF model which can distinguish between protein-coding sequences and intergenic (non-coding) regions. The NCBI genome repository was used to create training and validation data. GeneRFinder outperformed Prodigal and FragGeneScan in terms of specificity, with a score of 55% greater than Prodigal and 60% greater than FragGeneScan.

In [10], a meta gene caller (MGC) framework was proposed with additional amino acid usage to modify the Ophelia concept. Depending on the segment's GC content, the ORF is scored using the corresponding neural network model. The neural network approximated the chance of whether the ORF is coding or not. This work also evaluated MGC's performance on sequences with sequencing errors and compared results with other gene prediction methods. The MGC model achieved a best harmonic mean of 91.44% and a standard deviation of 0.15%. This concluded that MGC outperformed the state-of-the-art models in gene prediction.

A novel framework called CNN-MGP was proposed to predict genes without extracting features in [5]. It trained ten CNN models with ten datasets created based on the GC content. Candidate genes were ORFs with a likelihood of more than 0.5. CNN-MGP has a 94.87% average specificity, an 88.27% average sensitivity, and a 91.36% average harmonic mean. The harmonic mean's average standard deviation is 0.14%.

Prodigal [11] framework focused on three aims. Firstly, the recognition of the translation initiation site; secondly, the prediction of gene structure; and finally, reducing the false positives. This method used a dynamic programming algorithm for the prediction. When compared to the existing gene predictors, it achieved 96.7%.

An ML framework called FragGeneScan was proposed in [12]. This work utilized a hidden Markov model (HMM), which incorporated codon features with models that were designed to tackle sequencing errors to improve the performance of predicting protein-coding regions in short-length sequences. FragGeneScan was made up of two-level data abstraction representations. It used distinct states to indicate gene regions in the reverse and forward strands of a nucleotide sequence so that the genes can be predicted from both strands simultaneously. FragGeneScan achieved an average accuracy of 80%.

A novel framework called MetaGene was proposed in [13]. MetaGene combined di-codon frequencies calculated from a sequence's GC content with other measurements. The authors applied their model on the microbiome genomic sequences and done the annotations of genes. The model achieved 95.3% and 94% of sensitivity and specificity, respectively.

Balrog [14] is a prokaryotic gene-finding system that used a temporal convolutional network to learn patterns in protein sequences and predict gene locations. Balrog does not require genome-specific training. Even though Balrog found fewer genes than other gene predictors, it showed improved sensitivity. This concluded that Balrog could be used as an efficient tool for gene prediction.

A novel framework called virSearcher was proposed in [15], which used a CNN with gene information to identify phages in metagenomes, such as Bacteriophages that infect bacteria. Input sequences are encoded differently based on the positions of their coding and non-coding regions. Then, an embedding layer for word is used to convert the encoded sequences to word embeddings. Then, the CNN was used to identify the phage genome.

3 Methodology

Gene prediction is the task of predicting the probability of the given ORF sequence being coding or non-coding. The flow of the proposed framework is portrayed in Fig. 1.

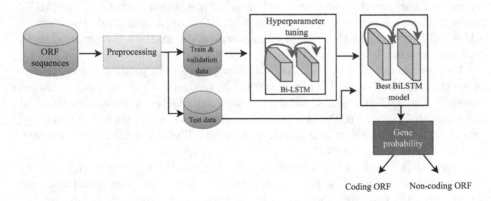

Fig. 1. Methodology of the proposed framework

3.1 Dataset

The dataset used in this work was used by the work called Orphelia by [7], which contains 7 million ORF sequences excised from 131 fully-sequenced prokaryotic genomes. These ORFs are extracted from DNA fragments of length 700 base pairs. GenBank [16] is used to annotate the genes in the sequences. The labels of the ORF sequences are binary, which indicates whether the ORF sequences are coding or non-coding. Due to limited computational power a subset of 100000 ORF sequences are used in this work. Out of the 100000 ORF sequences, 49975 sequences are coding and 50025 sequences are non-coding sequences.

3.2   Preprocessing

This step converts the ORF sequences to numerical values. The sequences and the respective labels in the data set are converted to binary values using one-hot encoding. It is a technique widely employed in bioinformatics to convert DNA sequences to numerical representation. Each nucleotide (A, T, G, C) in the DNA sequence is converted into a 4-dimensional numerical vector of 1 s and 0 s. Hence, we get a matrix for each sequence where the row number equals the sequence length and the column size equals four. It preserves the generality of the data because each nucleotide is uniquely represented. Figure 2 shows the encoding of an example sequence. The maximum length of the sequence in this dataset is 705. If the length of the sequence is less than 705, the sequence is padded with zero rows at the end. The entire set of sequences is divided into training, testing, and validation sets with a ratio of 70:15:15. The training set is given for model training, the validation set is utilized for hyperparameter tuning, and the test set is used to evaluate the model's performance. The flow of the preprocessing of the dataset is represented in the Fig. 3.

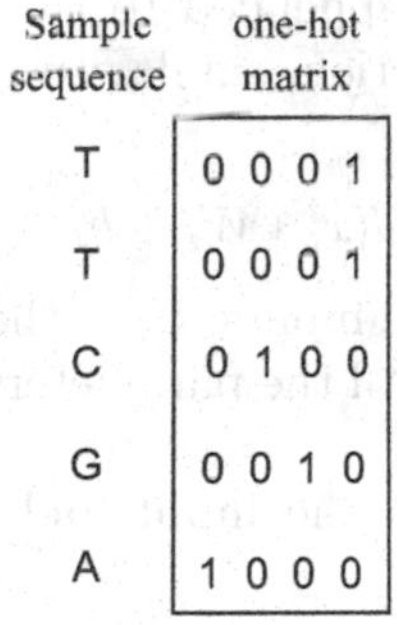

Sample sequence	one-hot matrix
T	0 0 0 1
T	0 0 0 1
C	0 1 0 0
G	0 0 1 0
A	1 0 0 0

Fig. 2. Example of one-hot encoding of a sample DNA sequence "TTCGA"

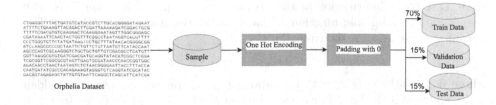

Fig. 3. Data Preprocessing

3.3 LSTM and Bi-LSTM

3.3.1 LSTM

LSTM networks address the vanishing gradient problem faced by RNN during training with long sequences. The gradients involved in the backpropagation become extremely small as they pass through multiple timesteps and the network layers. This makes the learning of long-term sequences challenging. The memory cells in the LSTM make it possible to process long-term dependencies in the sequences. Apart from that, LSTM has gates such as forget, input, and output. The gates are crucial in regulating information flow within the LSTM cells. Figure 4 shows the LSTM architecture, where x_t denotes the input signal and h_t is the output.

The forget gate determines what data should be erased, the input gate updates the memory cells, and the output gate decides the next concealed state. The hidden state holds the short-term memory of the previous output, which is given as input to the forget gate and input for that particular time step. The forget gate processes these inputs by multiplying with weights and adding bias. Then, a sigmoid function is applied. If the sigmoid output is 0, the model forgets everything; if the output is 1, the current output is multiplied by the previous cell state. The Eq. 1 depicts the forget gate's operation.

$$f_t = sigmoid(x_t * W_f + h_{t-1} * U_f + b_f) \tag{1}$$

where W_f is the learnable weights correlated with the inputs, h_{t-1} denotes the previous hidden state, U_f is the weights associated with the hidden state, and b_f denotes the bias. All the above parameters are associated with the forget gate.

The amount of information appended to the cell state in the input gate is regulated using the sigmoid function, similar to the forget gate. The input gate operation is shown in Eq. 2.

$$i_t = sigmoid(x_t * W_i + h_{t-1} * U_i + b_i) \tag{2}$$

where W_i is the weights with the inputs, U_i is the weights corresponding to the hidden state, and b_i is the bias. All the parameters are associated with the input gate.

By adding the outputs from the input and forget gates, the cell state is updated as illustrated in Eq. 3.

$$c_t = f_t * c_{t-1} + i_t * tanh(x_t * W_c + h_{t-1} * U_c + b_c) \tag{3}$$

where W_c is the weights with the inputs , U_c is the weights with the hidden state, c_{t-1} is the information at the previous time step and b_c is the bias. The output gate uses sigmoid function similar to other two gates for regulation. The output gate equation is shown in Eq. 4.

$$o_t = sigmoid(x_t * W_out + h_{t-1} * U_out + b_out) \tag{4}$$

where W_out is the weights with the inputs, U_out is the weights with the hidden state, and b_out is the bias. The current output h_t of LSTM is calculated by Eq. 5.

$$h_t = o_t * tanh(c_t) \tag{5}$$

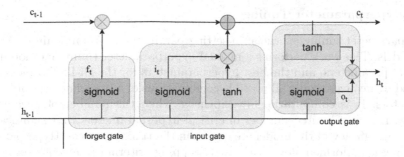

Fig. 4. Architecture of LSTM

3.3.2 Bi-LSTM

It has the input flow in forward and backward directions. It helps to preserve the future and the past information. It is like two LSTM models working on the same sequence in different directions. The following Fig. 5 shows the overall structure of Bi-LSTM.

The architecture of Bi-LSTM consists of two separate LSTM networks. One LSTM network takes the input sequence as it is, while the other takes it in reverse order. The two LSTM models can have two separate hidden layers and then feed forward to the same output layer. Both LSTM models will give the probability as outputs. These outputs are combined to get the final output as shown in Eq. 6.

$$h_t = h_t^f \oplus h_t^b \tag{6}$$

where h_t is the final probability vector of the model, h_t^f denotes the output probability from the LSTM model in the forward direction, and h_t^b is the output probability from the opposite LSTM. $\oplus$ indicates the concatenation operation.

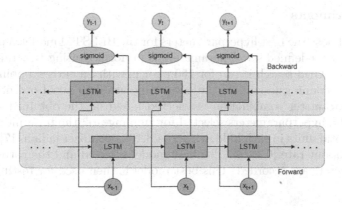

Fig. 5. Architecture of Bi-LSTM

3.4 Hyperparameter Tuning

Hyperparameter tuning is crucial in setting parameter values in training ML or DL models. This work performs a manual hyperparameter tuning method using a grid of parameters, and the best hyperparameters for the Bi-LSTM model are selected. It involves trying all possible combinations of a predefined set of hyperparameters. It is computationally expensive when the number of parameters is more. Hence, a limited number of critical hyperparameters that significantly modify the efficacy of the model are chosen in the tuning process. Hyperparameter tuning is performed along with early stopping criteria to prevent overfitting. Early stopping monitors the model's validation loss and halts the training process when a specific condition is met. The condition used is the patience value, which is the number of epochs for which the validation loss is not improving or getting worse. The best model obtained during the hyperparameter tuning is used for the final prediction.

4 Experimental Setup and Evaluation Metrics

This section presents the software and hardware settings, the experiments conducted in this work, and the metrics used to evaluate the model.

4.1 Implementation Details

Python3 is used to code this project using jupyter Notebook environment. The main libraries utilized are Keras API in TensorFlow version 2.9.2, and Sklearn version 1.0.2. The computations are done by machine eauipped with 16 GB RAM, Intel i7 processor and a NVIDIA GEFORCE GPU with 4 GB RAM running on the Ubuntu.

4.2 Experiments

In order to choose the best hyperparameters for the Bi-LSTM model, a parameter tuning with a predefined set of parameters is performed using the train set and validation set. Grid search is run for 150 epochs with an early stopping criteria based on the training loss and a patience value as 10 to avoid overfitting. The best hyperparameter combination is selected based on the obtained validation accuracy. The hyperparameters selected for tunning are: the total hidden layers [1,2,3], the total units in each hidden layer [4,8,16], optimizer [adam [17], adagrad [18]] and dropout rate [0.3,0.4]. The best model is saved and the corresponding hyperparameters are recorded. This best model is then used for testing.

A 1-dimensional CNN with two convolutional layers and an LSTM model with one layer is selected for the comparison experiment. After each convolution layer, a max pooling layer of pool size 2 exists. The convolutional layers use a kernel size of 21, and the total filters are 64 and 200, respectively, for the first and second layers. Then, a dropout layer with 0.5 as the dropout percentage is used. Followed by this is a dense layer with 128 neurons. Again, a dropout layer is added, and its output is given to the output layer with two neurons. The LSTM model has 16 units in one layer, then a dropout layer is added with a 0.2 dropout rate, and finally, a dense layer with 128 neurons.

The proposed Bi-LSTM model is compared with the CNN-MGP proposed in [5]. For comparison, CNN-MGP is trained with the training set. The testing data is used to estimate the performance of the model.

4.3   Evaluation Metrics

Statistical measures such as Accuracy, AUC, Precision, Recall and F1-score are measured to evaluate the model efficiency. Let True Positive (TP) and True Negative (TN) represent the number of correctly classified coding and non-coding ORFs, respectively. Let False Positive (FP) and False Negative (FN) represent the number of incorrectly classified coding and non-coding ORFs, respectively. Accuracy measures the overall correctness of the model. AUC calculates the area under the curve plotted between TP and FP values by varying the classification threshold. Precision is measured as the percentage of correct positive predictions among all positive predictions made. Recall calculates the model's ability to identify all instances of the coding class. F1-score combines both Precision and Recall. It is useful when the dataset is imbalanced.

5   Results and Discussion

The findings of hyperparameter tuning are showcased in Tables 1, 2 and 3. From the table, it could be observed that the best validation accuracy of 0.9548 is obtained with two hidden layers, adam optimizer, a dropout rate of 0.3 and the units in the first and the subsequent hidden layers as 8 and 16, respectively. With this hyperparameter setting, the Bi-LSTM model is able to converge at an epoch of 130 and achieve the best validation accuracy.

Table 1. Findings of hyperparameter tuning with hidden layer number = 1

No. of units in the hidden layer	Optimizer	Learning rate	Validation Accuracy
4	adam	0.3	0.9054
8	adam	0.3	0.9409
16	adam	0.3	0.9073
4	adam	0.4	0.9173
8	adam	0.4	0.9102
16	adam	0.4	0.9213
4	adagrad	0.3	0.8501
8	adagrad	0.3	0.8389
16	adagrad	0.3	0.8673
4	adagrad	0.4	0.8728
8	adagrad	0.4	0.8802
16	adagrad	0.4	0.8780

Table 2. Findings of hyperparameter tuning with hidden layer number = 2

No. of units in the first hidden layer	No. of units in the second hidden layer	Optimizer	Learning rate	Validation Accuracy
4	8	adam	0.3	0.9246
8	**16**	**adam**	**0.3**	**0.9548**
16	32	adam	0.3	0.9264
4	8	adam	0.4	0.9201
8	16	adam	0.4	0.9245
16	32	adam	0.4	0.9189
4	8	adagrad	0.3	0.8471
8	16	adagrad	0.3	0.8663
16	32	adagrad	0.3	0.8682
4	8	adagrad	0.4	0.839
8	16	adagrad	0.4	0.8649
16	32	adagrad	0.4	0.8694

The model with the best performance is saved and used for final evaluation. The results achieved by the best Bi-LSTM model is presented and compared against the CNN and LSTM models in Table 4. Analysis of the table explains that the Bi-LSTM model is the best performing model. The CNN outperforms the LSTM model. However, Bi-LSTM gives promising results like 0.99 AUC, 0.9537

Table 3. Findings of hyperparameter tuning with hidden layer number = 3

No. of units in the first hidden layer	No. of units in the second hidden layer	No. of units in the third hidden layer	Optimizer	Learning rate	Validation Accuracy
4	8	16	adam	0.3	0.9129
8	16	32	adam	0.3	0.9164
16	32	64	adam	0.3	0.9293
4	8	16	adam	0.4	0.9124
8	16	32	adam	0.4	0.9171
16	32	64	adam	0.4	0.919
4	8	16	adagrad	0.3	0.861
8	16	32	adagrad	0.3	0.8727
16	32	64	adagrad	0.3	0.8699
4	8	16	adagrad	0.4	0.8575
8	16	32	adagrad	0.4	0.8623
16	32	64	adagrad	0.4	0.8726

Accuracy, 0.9653 Precision, 0.9457 Recall and 0.9522 F1-score. The improved results of Bi-LSTM show that it can capture more context and long-term dependencies in the input data than LSTM and CNN. Because, Bi-LSTM processes the input sequence in both forward and backward directions.

Table 4. Comparison of the results achieved by the best Bi-LSTM and the CNN, LSTM model with test data (The best results are highlighted)

Model	AUC	Accuracy	Precision	Recall	F1-score
LSTM	0.9785	0.9283	0.9285	0.9279	0.9282
CNN	0.9878	0.9532	0.9602	0.9452	0.9518
Bi-LSTM	**0.9901**	**0.9537**	**0.9653**	**0.9457**	**0.9522**

The compaison results of analysis of the proposed model with CNN-MGP is presented in Table 5. It can be observed from the findings that the Bi-LSTM model outperforms the CNN-MGP model.

Table 5. Comparison of the results achieved by the best Bi-LSTM and the CNN-MGP model with test data (The best results are highlighted)

Model	AUC	Accuracy	Precision	Recall	F1-score
CNN-MGP	0.99	0.9427	0.9523	0.9378	0.9449
Bi-LSTM	**0.9901**	**0.9537**	**0.9653**	**0.9457**	**0.9522**

6   Conclusion

Recently, significant focus has been placed on applying DL methods to various bioinformatics problems. This study uses a Bi-LSTM model to classify ORF sequences into protein-coding ORF sequences or non-coding sequences. Bi-LSTMs have been used for many applications. This work starts with pre-processing the ORF sequences using one-hot encoding. The one-hot encoded sequences are then passed to the Bi-LSTM layers. The final dense layer generates the likelihood that an ORF encodes a gene. The hyperparameter tuning identifies the best parameters for the proposed model. The best model is used to evaluate the gene prediction framework. The proposed model produces promising results than a CNN and LSTM model.

The duration of the hyperparameter tuning is a limitation of this work. Hence, using transformer networks for classification can be considered in the future because transformers are particularly well-suited for parallelization due to their architecture. Additionally, adding more genomes of different varieties of microorganisms to increase the dataset size can be considered as future work.

References

1. Thomas, T., Gilbert, J., Meyer, F.: Metagenomics-a guide from sampling to data analysis. Microb. Inf. Exp. **2**, 1–12 (2012)
2. Wang, Z., Chen, Y., Li, Y.: A brief review of computational gene prediction methods. Genomics Proteomics Bioinf. **2**(4), 216–221 (2004)
3. Sharpton, T.J.: An introduction to the analysis of shotgun metagenomic data. Front. Plant Sci. **5**, 209 (2014)
4. LeCun, Y., Bengio, Y., Hinton, G.: Deep learning. Nature **521**(7553), 436–444 (2015)
5. Al-Ajlan, A., El Allali, A.: CNN-MGP: convolutional neural networks for metagenomics gene prediction. Interdisc. Sci. Comput. Life Sci. **11**, 628–635 (2019)
6. Arango-Argoty, G., Garner, E., Pruden, A., Heath, L.S., Vikesland, P., Zhang, L.: Deeparg: a deep learning approach for predicting antibiotic resistance genes from metagenomic data. Microbiome **6**, 1–15 (2018). https://doi.org/10.1186/s40168-018-0401-z
7. Hoff, K.J., Tech, M., Lingner, T., Daniel, R., Morgenstern, B., Meinicke, P.: Gene prediction in metagenomic fragments: a large scale machine learning approach. BMC Bioinf. **9**, 1–14 (2008)
8. Al-Ajlan, A., El Allali, A.: Feature selection for gene prediction in metagenomic fragments. BioData Min. **11**(1), 1–12 (2018)
9. Silva, R., Padovani, K., Góes, F., Alves, R.: geneRFinder: gene finding in distinct metagenomic data complexities. BMC Bioinf. **22**(1), 1–17 (2021)
10. El Allali, A., Rose, J.R.: MGC: a metagenomic gene caller. BMC Bioinf. **14**, 1–10 (2013)
11. Hyatt, D., Chen, G.-L., LoCascio, P.F., Land, M.L., Larimer, F.W., Hauser, L.J.: Prodigal: prokaryotic gene recognition and translation initiation site identification. BMC Bioinf. **11**, 1–11 (2010)
12. Rho, M., Tang, H., Ye, Y.: FragGenesSan: predicting genes in short and error-prone reads. Nucleic Acids Res. **38**(20), e191–e191 (2010)

13. Noguchi, H., Park, J., Takagi, T.: Metagene: prokaryotic gene finding from environmental genome shotgun sequences. Nucleic Acids Res. **34**(19), 5623–5630 (2006)
14. Sommer, M.J., Salzberg, S.L.: Balrog: a universal protein model for prokaryotic gene prediction. PLoS Comput. Biol. **17**(2), e1008727 (2021)
15. Liu, Q., et al.: virSearcher: identifying bacteriophages from metagenomes by combining convolutional neural network and gene information. IEEE/ACM Trans. Comput. Biol. Bioinf. **20**(1), 763–774 (2022)
16. Benson, D.A., et al.: Genbank. Nucleic Acids Res. **41**(D1), D36–D42 (2012)
17. Kingma, D.P., Ba, J.: Adam: a method for stochastic optimization, arXiv preprint arXiv:1412.6980 (2014)
18. Duchi, J., Hazan, E., Singer, Y.: Adaptive subgradient methods for online learning and stochastic optimization. J. Mach. Learn. Res. **12**(7), 2121–2159 (2011)

Energy-Efficient Task Scheduling in Fog Environment Using TOPSIS

Sukhvinder Singh Nathawat[✉] and Ritu Garg

Computer Engineering Department, NIT Kurukshetra, Kurukshetra, Haryana, India
{sukhvinder_32113205,ritu.59}@nitkkr.ac.in

Abstract. As IoT devices proliferate quickly, the amount of data generated has increased rapidly, requiring analysis and storage in cloud data centers. To mitigate high data traffic and reduce latency, fog emerged as a paradigm that brings cloud services closer to users through accessible networks. By doing so, fog computing alleviates traffic congestion and delays. Moreover, fog devices are constrained in terms of power supply, processing capabilities, and communication resources, making it challenging to design fog systems that meet real-time application requirements. Fog devices are having limited power supply because most of them are battery operated, thus energy conservation becomes a crucial objective. Energy-efficient task scheduling is one strategy to reduce energy consumption in fog systems. However, finding an efficient scheduling strategy that balances system performance, energy consumption and meets service level agreements remains a challenge. Therefore, to address these issues, in this paper, we proposed an energy efficient scheduling algorithm using a multi-criteria approach i.e. Technique for Order of Preference by Similarity to Ideal Solution (TOPSIS) to optimize energy, execution performance and cost in fog computing. To evaluate the proposed approach, we conducted simulations in MATLAB and the results demonstrate its effectiveness in significantly reducing makespan, cost and energy usage in the fog computing environment.

1 Introduction

The proliferation of IoT has led to a remarkable surge in the number of connected devices. According to projections from the statistics portals, the global count of connected devices is estimated to touch 76.55 billion by 2026 [1, 2]. Data generated by these devices is huge, which is sensitive to distance between the source and the consumer (i.e., cloud). Moreover, it also increases the network traffic over the cloud as IoT devices generate data very frequently which results in degradation of execution performance. Thus, to tackle this issue a new computing paradigm was introduced by Cisco in 2014 i.e., fog computing [3, 4]. Fog is an extension to the cloud as it brings computing services nearer to end devices.

Fog computing is being widely utilized across various domains, including video surveillance, smart cities, and transportation. It facilitates the establishment of distributed networks consisting of computing and networking resources that can be strategically deployed based on their proximity to where they are most needed [4]. Unlike traditional

K. K. Patel et al. (Eds.): icSoftComp 2023, CCIS 2031, pp. 16–28, 2024.
https://doi.org/10.1007/978-3-031-53728-8_2

centralized systems, fog computing positions a fog layer in close proximity to end devices, effectively bridges the gap between these devices and the cloud. However, it is resource constrained in nature as compared to cloud in terms of energy, computation abilities as well as storage.

As fog is a subset of cloud it also suffers from various issues that occur in cloud, but we need different approaches to tackle them due the constrained nature of fog. The major issues in fog computing are energy consumption, cost, security, scalability, execution time etc. Among these energy consumption is regarded as the most significant one as fog devices do not have uninterrupted power supply, they mostly run over batteries. Therefore, optimizing the energy consumption is very essential as it helps in increasing the life of the device [4, 5].

Major portion of the energy is consumed in the computation and if we can optimize it then it will increase the performance of fog system. By decreasing the consumption of energy of the fog nodes, it leads to cut down expenses and increases the lifetime of the fog nodes. Task scheduling is one of the effective solution to address the constraints. Task scheduling refers to the process of efficiently allocating the computing resources in a fog environment. Task scheduling lies in the category of the NP-hard problem. So heuristic approaches are best possible solutions. In literature, state of art heuristics like HEFT [5], Min-Max [7], Min-Min [8] etc. However, they do not deal with multiple objective simultaneously. However, issue of task scheduling in fog system is multi-objective in nature.

Thus, in this article, we have used a heuristic technique i.e. TOPSIS for scheduling the tasks on the fog to optimize multiple conflicting objectives of cost, time of execution as well as energy usage. TOPSIS accommodates decision-making scenarios that involve multiple criteria. It enables the evaluation of alternatives based on different attributes or dimensions simultaneously, making it suitable for complex decision problems. Thus, TOPSIS is consider as one of the most suitable heuristic approach for generating task scheduling solution with an objective to optimize cost, time as well as energy simultaneously. Therefore, in this paper, we have utilized the TOPSIS for energy saving as a primary concern while minimizing the execution time (makespan), cost. The details of our contributions are provided below:

- In this paper, to solve of independent task scheduling issue in fog computing as a multi-objective optimization issue, we have taken energy consumption, cost and makespan as optimization parameters.
- We presented a task scheduling solution using TOPSIS method in fog computing as TOPSIS accommodates decision-making scenarios involving multiple criteria and provide the best potential solution.
- To demonstrate the effectiveness of the developed technique, the simulation findings are compared with three widely recognized existing algorithms PSO [6], MinMax [7], MinMin [8]. This comparative analysis serves as a means to showcase the superiority and performance of the proposed algorithm.

The remainders of the paper are as follow: Sect. 2 provides a comprehensive examination of various task scheduling techniques that have been developed. Section 3 address the system architecture, the problem formulation that has been proposed, and the technique that will be employed. The result and analysis is done in Sect. 4. Ultimately, the present study concludes in Sect. 5, wherein we draw our final conclusions.

2 Work Done

This section delivers an overview of the work that conducted in the field of fog computing, specifically focusing on efficient task scheduling. Task scheduling in fog computing has been the focus of numerous studies and research endeavors. These researchers employ various methods and approaches to effectively allocate and schedule tasks among fog nodes. The diversity among these researchers lies in their distinct methodologies and techniques employed for task scheduling in fog environments. Some examples of related work in this area includes:

In Ref. [9] authors have presented a priority based task scheduling technique for enhancing performance of the fog. In this technique, the authors have assigned the priority to the incoming task based upon their criticality, and then these tasks are buffered in the request. And as soon as the best suited fog node is available, the task is assigned to it. The results depict the improvement in the performance of the application.

In Ref. [10] authors have introduced a meta-heuristic technique known as the Bees Life Algorithm (BLA) with the objective of addressing scheduling issues within the context of fog system. The motive of this study is to determine an optimal CPU execution time by minimizing the allocation of tasks across all fog nodes. Similarly, in [11] authors have presented a task scheduling method in a hybrid cloud-fog environment, by utilizing the concept of a genetic algorithm (GA). The motive of this technique is to minimize the overall latency associated with service requests. However, the authors of the aforementioned techniques considering only a single objective for task scheduling in fog system.

To improve performance of the fog nodes, in [12] the authors have developed a hybrid task scheduling technique that consists of TOPSIS for ranking the fog nodes and Analytic Hierarchy Process for calculating the priority weight of the fog nodes. This technique was developed to find the best suited fog node for assigning the task with the aim to ensure performance along with cost.

Similarly, in Ref [13] the authors have presented an algorithm for scheduling namely, Energy Efficient Task Scheduling in Fog Computing. The presented approach is based on the Particle Swarm Optimization with an aim to optimize the makespan, energy as well as execution time. Additionally, they have not considered the cost as a performance metric parameter.

As per SLAs, economic cost is another key parameter to be addressed. The authors in [14] suggested a multi-objectives grey wolf optimizer to allocate the tasks to the fog nodes. They have also utilized the concept of crowding distance method to ensure optimal utilization of the energy as well as to reduce the overall computational cost.

Due to growing demands of energy aware or green computing solutions especially in heterogeneous computing systems like fog, it is desirable to explore the approaches to solve the scheduling problem. Therefore, to tackle the observed shortcomings i.e., optimal energy consumption, execution performance and economic cost, we have developed an energy efficient task scheduling technique utilizing the TOPSIS. Our proposed technique ensures minimum energy usage, good performance and reduction in cost.

3 Proposed Work

3.1 System Architecture

We are considering the three-tier architecture consisting of the cloud, fog and end devices. The fog layer comprising fog nodes deployed closer to the edge, acts as an intermediary, providing storage as well as computational capabilities. It receives tasks from sensors/end devices and schedules them based on factors like cost, available resources, priority and energy consumption. There exists a fog administration that plays a crucial role in allocating tasks efficiently. In this work, our prime focus is on fog layer in which fog administration's role is vital in achieving energy efficiency, improving system performance and reducing costs. Figure 2 demonstrates proposed energy-efficient task scheduling framework in fog network.

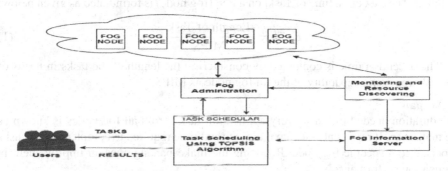

Fig. 2. Task Scheduling Framework for Fog Environment

End users send tasks for execution to the task scheduler. Task scheduler receives incoming tasks or jobs and make decisions about their mapping\scheduling using proposed algorithm based on predefined criteria such as cost, usage of energy, execution time etc. Fog administrator plays a vital part in assigning tasks to the fog nodes as per scheduling decision of the task scheduler.

Further, the Monitoring and Resource Discovery module is responsible for discovering and monitoring fog resources, continuously collecting information about the available fog resources as well as actively discovering new resources. Additionally, it transmits the status information to the Fog Information Server (FIS) which is close to the Task Scheduler. As a result, FIS maintains an up-to-date and comprehensive database regarding the availability of fog resources and shares information to the task scheduler.

3.2 Problem Formulation

In a fog computing environment, tasks which are latency sensitive or light weight are executed on the fog nodes and rest are carried out on cloud computing. The problem of task scheduling is the efficient allocation of tasks to resources considering various constraints and objectives. In this paper, we have considered the energy as primary optimization objective along with secondary objective i.e. cost and execution time.

In the fog, tasks can be modeled as $Task_i \varepsilon \{Task_1, Task_2, \ldots, Task_m\}$. A task has characteristics like memory required, number of instructions, how vast the input and output files are. Given a set of tasks with different characteristics will be scheduled on the various fog nodes that are present in the fog environment. $FN_j \varepsilon \{FN_1, FN_2 \ldots FN_p\}$ represents the set of fog nodes, each fog node having varying processing capabilities, frequency etc. The goal is to create an efficient scheduling that assigns tasks to different fog nodes in order to achieve the following objectives: cost, execution time, and energy usage.

Execution Time
In the context of task scheduling, execution time (EXT) for a task on a fog node refers to the amount of time needed by the fog node to finish the execution of a particular task assigned. The execution time of $Task_i$ on a FN_j (fog node) is formulated as given below:

$$EXT_i^j = \frac{\text{Length of Task}}{\text{MIPS}}. \tag{1}$$

The execution time is computed by considering the length of the tasks in terms of MI and processing capability of the jth fog node as MIPS.

Makespan
The duration needed to finish every application task across all fog nodes is known as the makespan. It represents the maximum of total time required to complete entire tasks allocated to a specific fog node. Reducing the makespan signifies an improvement in application performance:

$$\text{Minimize Makespan} = \text{Max}\left(\text{Total\_EXT}^j\right). \tag{2}$$

where $Total\_EXT^j$ is the total time of execution of all tasks assigned to fog node FN_j..

Energy Consumption
The quantity of energy the fog nodes use during task execution and operations in fog is referred to as energy consumption. The energy consumption of the $Task_i$ on a fog node FNj can be computed as:

$$\text{Energy\_Used}\left(EU_i^j\right) = A * C * V^2 * f \tag{3}$$

The energy is computed by considering the voltage level V and operating frequency f of a fog node FN_j whereas the coefficient C and A are the values of the total capacitance as well as number of the switching activity per clock cycle respectively. The main goal is to reduce the amount of energy used by all the tasks, which is calculated as:

$$\text{Minimize Total Energy\_Used} = \sum \text{Energy\_Used}\left(FN_j\right). \tag{4}$$

Cost

In fog computing, the cost of task scheduling refers to the expenses associated to the execution of tasks and the overall operation. The cost in order to executing the $Task_i$ on FN_j is computed as:

$$Cost_i^j = Price_j^* EXT_i^j \tag{5}$$

where $Price_j$ is the cost of the jth fog node per unit time interval. The key motive is to optimize the total execution cost of tasks on fog nodes, which is displayed by:

$$\text{Minimize Cost} = \sum Cost_i^j \tag{6}$$

3.3 TOPSIS Algorithm

The TOPSIS approach was initially introduced by Yoon and Hwang [15] as a multi-criteria decision-making approach. According to TOPSIS, the taken alternatives must have the greatest geometric distance from the negative ideal solution and closest geometric distance to the positive ideal solution. It uses the Euclidean distance to calculate how close an alternative to the ideal answer is. The best achievable values for each attribute are combined to form the positive ideal solution, while worst values for each attribute are included in negative ideal solution. To determine the comparative proximity to positive ideal solution, TOPSIS calculates the distances to both the negative ideal solution and the positive ideal solution. By comparing these relative distances, a prioritized order for the alternatives can be established. Therefore, TOPSIS is a valuable technique in multi-criteria decision-making, where it assesses the alternatives based on their proximity to distance from the positive ideal solution to the negative ideal solution. Due to its computational efficiency, simplicity, comprehensibility as well as ability to gauge the relative performance of decision alternatives; this method is widely employed in decision-making processes.

The steps involved in the TOPSIS method are as shown:

1. *Define the decision matrix:* Make a decision matrix that depicts how well each fog node FN_j performs in relation to each criterion (C_r) i.e. energy used, cost and execution time. Each row represents an alternative, and each column represents a criterion.
2. *Normalize the decision matrix:* To make certain that every criterion is on a comparable scale, normalize the decision matrix. The usual method for doing this is to divide every element in a column by the square root of the total of the squares of all the components in that column.

$$N_j = \sqrt{\sum C_{ij}^2} \tag{7}$$

$$C_{ij} = \frac{C_{ij}}{N_j} \tag{8}$$

Table 1. Decision Matrix for proposed Work

$$
\begin{array}{c}
 \quad C_1 \qquad C_2 \qquad C_3 \\
\begin{array}{c} FN_1 \\ FN_2 \\ \vdots \\ FN_p \end{array}
\begin{bmatrix}
EXT_i^j & EU_i^j & Cost_i^j \\
EXT_i^j & EU_i^j & Cost_i^j \\
EXT_i^j & EU_i^j & Cost_i^j \\
EXT_i^j & EU_i^j & Cost_i^j
\end{bmatrix}
\end{array}
$$

3. *Assign weights to criteria:* By allocating weights, we define the relative relevance of each criterion. The importance of each criterion in the decision-making process is reflected in the weights. The sum of all weights should be equal to 1.

$$\sum_1^n w_i = 1 \tag{9}$$

4. *Calculate the weighted normalized decision matrix:* Multiply each normalized value in the decision matrix by the weight (w_i) that corresponds to it to generate a weighted normalized decision matrix.

$$\text{Weighted Normalized DM} = c_{ij*}\, w_i \tag{10}$$

5. *Compute I^+ (positive ideal solution) and I^- (negative ideal solution):* For each criterion, identify the best and worst values among all alternatives. The highest values for each criterion are symbolized by I^+, and the minimum values are by I^-.

   For I^+: Identify the maximum value in column j:

$$I^+ = \max(C_{ij}) \tag{11}$$

   For I^-: Identify the minimum value in column j

$$I^- = \min(C_{ij}) \tag{12}$$

6. *Calculate the Euclidean distances:* Calculate the Euclidean distance among every alternative as well as the negative and positive ideal solutions. The Euclidean distance is computed depending on the weighted normalized decision matrix.

   Euclidean distance between alternative i and positive ideal solution I^+:

$$D_i^+ = \sqrt{\left(\sum (R_{ij} - I_j^+)^2\right)} \tag{13}$$

   Calculate Euclidean distance between p and I^-I^-

$$D_i^- = \sqrt{\left(\sum (R_{ij} - I_j^-)^2\right)} \tag{14}$$

7. *Compute the relative closeness:* The relative proximity of each alternative can be ascertained by dividing the distance by the total of the distances to both negative ideal solutions and positive, to negative ideal solution. Compute the relative closeness coefficient R C_i as:

$$RC_i = \frac{D_i^-}{(D_i^+ + D_i^-)} \tag{15}$$

8. *Rank the alternatives:* Rank the alternatives based on their relative closeness values. The alternatives are sorted in descending order of their R C_i values. The alternative with the highest relative closeness is considered the most preferred or the best solution.

The TOPSIS algorithm provides a systematic and quantitative approach to decision-making in situations with multiple criteria. In this paper we have considered the three criteria i.e. energy consumption, cost and the execution time.

Algorithm 1: TOPSIS

Input: Decision matrix DM_i with p fog node and r criteria, weight vector, $W_i = \{w_1, w_2 \ldots w_r\}$ representing the relative importance of each criterion.

```
for j= 1to p:
   for k= 1 to r:
      Determine the Normalized DM   //Using eq. (7) and (8)
   end for
end for

for j=1to p:
   for c= 1 to r:
      Compute weighted Normalized DM        //Using eq. (10)
   end for
end for

Calculate I⁺ and I⁻                 //using eq. (11) and (12)

for j=1 to p:
   for c=1 to r:
      Calculate Euclidean distance between p and I⁺
                                    //using eq. (13)
   Calculate Euclidean distance between p and I⁻
                                    //using eq. (14)
   end for
end for

for j=1 to p:
   Compute RCᵢ                      //using eq. (15)
end for
Rank the alternatives based on their relative closeness
values
return highest RCᵢ
```

The algorithm offers a methodical framework for assessing and prioritizing alternatives by considering their proximity and their distance from the positive ideal solution to the negative ideal solution the output is a ranking of the alternatives, providing insights into the most suitable options according to the given criteria and their relative importance.

3.4 Proposed Task Scheduling Algorithm Using TOPSIS

A detailed description of the presented algorithm for task scheduling in fog environment to optimize energy consumption, cost, as well as execution time is presented in this section. We have utilized the concept of the TOPSIS algorithm to schedule the task on fog nodes. By applying the TOPSIS algorithm to task scheduling in fog system, the resulting ranking of fog nodes can guide the allocation of tasks to fog nodes, taking into account the optimization objectives of the cost, energy consumption, execution time.

Algorithm 2: Task Scheduling Using TOPSIS

Input: set of tasks (m), set of fog node (p), criteria (C_r) i.e. energy consumed, execution time, cost.
Output: Optimal Solution

```
for i=1 to m
        for j=1 to p
        Compute Execution Time (EXT_i^j) // Using eq. (1)
        Compute Energy Consumption (EU_i^j)// Using eq. (3)
        Compute Cost (Cost_i^j)           // Using eq. (5)
        end for
end for

For each task ∈ Task_i

        for j=1 to p
            for c=1 to r

            Create   a   decision   matrix   DM   with   row
        representing   fog   node   and   columns   representing
        optimization     objective/criteria     C_r     (energy
        consumption, cost, execution time).

                //as mentioned in Table 1.

        S← Call TOPSIS (DM)

            end for

        end for

    Return Schedule S //Where 'S' is a Scheduling Solution
    Return Total Energy(S), Total Cost (S), Makespan (S)
```

4 Results and Analysis

The details of the simulation setup also investigation of the findings of the suggested method are discussed in this section. The evaluation as well as the comparison of our proposed algorithm are based on parameters: energy consumption makespan, and cost. To benchmark performance of the presented method, we compare it with three existing algorithms namely PSO [7], MinMin [8] and MinMax [9].

For executing the proposed approach, we utilized the MATLAB software, which is a powerful programming platform commonly used for system design and analysis. Since fog nodes in our fog infrastructure possess varying processing power, our study involves the assumption that every individual node possesses a distinct processing capacity, quantified in MIPS (Millions of Instructions per Second). Additionally, we considered the supporting voltage levels (Volts) and operating frequency (GHz) for each fog node. To replicate a realistic fog infrastructure, we included twelve processing nodes or fog nodes in our simulation, each with its own unique characteristics with a different number of tasks i.e. 20,50,100 (the details of the same is provided in Table 2).

The parameters w_i (weight value) used by the TOPSIS is initial as $w_1 = 0.4$ (for EU), $w_2 = 0.3$ (EXT_i^j),$w_3 = 0.3$ $(Cost_i^j)$. The sum of the values of the w_i must be equal to one i.e. $\sum w_i = 1$. Our primary objective in this article is energy saving thus weight of the EU considered is slightly higher than cost and makespan.

Table 2. System Model Characteristics.

Parameters	Values	Unit
Number of the fog nodes	12	
Number of the Tasks	20/50/100	
Voltage	[1.1, 1.6]	Volt
Processing rate	[1200,1600]	MIPS
Frequency	[0.9, 1.2]	GHz
Number of Instructions	[1,100]	109 instruction

4.1 Analysis of Energy Consumption

Energy consumption metric gives the energy used to execute tasks in fog nodes. The primary motive of this paper is to optimize the energy used for fog because fog devices are resource constrained devices and mainly battery based, hence energy saving becomes an important concern in fog. The energy used by $F N_j$ is computed using Eq. (3). The obtained readings of energy used corresponding to the proposed algorithm and other considered algorithms are shown in Fig. 3 and the findings depicts that the presented algorithm outperformed in terms of energy saving.

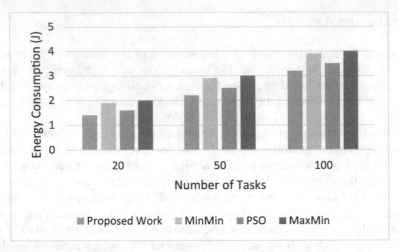

Fig. 3. Energy Used at Different Number of Tasks

4.2 Analysis of Makespan

The makespan of the proposed algorithm is equated with some heuristic/meta-heuristic techniques. The makespan is computed with the help of the Eq. (2). Figure 4 demonstrate that the presented approach provide better results in terms of makespan than PSO, Min-Min and Max-Min.

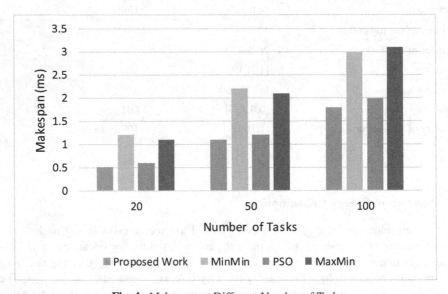

Fig. 4. Makespan at Different Number of Tasks

4.3 Analysis of Cost

The cost metric quantifies the total expenditure incurred for executing the tasks on the fog nodes. As cost is also an effective parameter in fog, optimization of cost leads to better QoS to the users as per SLA. The Fig. 5 demonstrates the comparative analysis for cost with respect to the number of the tasks. The cost is determined using Eq. (5). The obtained readings of the presented algorithm and other considered algorithms are shown in Fig. 5 and the findings indicate that the algorithm proposed in this study demonstrated greater efficiency in terms of cost.

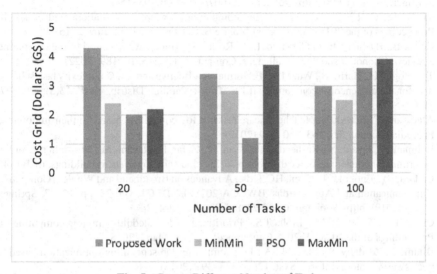

Fig. 5. Cost at Different Number of Tasks

5 Conclusion

Energy conservation is a crucial issue in fog systems due to their limited power supply. Excessive energy consumption not only leads to higher operational costs but also contributes to environmental harm through increased greenhouse gas emissions. Thus energy savings is an import issue to be addressed. Further, to enhance the efficiency of fog systems, the implementation of an efficient task scheduler becomes essential. This paper presents a novel task-scheduling method using TOPSIS to optimize the energy used, cost and makespan in fog computing. The TOPSIS algorithm provides a systematic and quantitative approach to decision-making in situations with multiple criteria. To conduct an assessment to evaluate the effectiveness of the presented task scheduling algorithm, we performed the simulations utilizing MATLAB. We compared its performance against other widely used heuristics and meta-heuristic techniques such as Min-Min, PSO, and Max-Min. The findings illustrate that the algorithm proposed in this study outperformed the other techniques in terms of energy saving, cost as well as makespan.

References

1. Kunal, S., Saha, A., Amin, R.: An overview of cloud-fog computing: architectures, applications with security challenges. Secur. Priv. **2**(4), e72 (2019)
2. Mukherjee, M., et al.: Security and privacy in fog computing: challenges. IEEE Access **5**, 19293–19304 (2017)
3. Hazra, A., Rana, P., Adhikari, M., Amgoth, T.: Fog computing for next-generation internet of things: fundamental, state-of-the-art and research challenges. Comput. Sci. Rev. **48**(100549), 4 (2023)
4. Michalewicz, Z.: Genetic Algorithms + Data Structures = Evolution Programs, 3rd edn. Springer, Berlin (1996). https://doi.org/10.1007/978-3-662-03315-9
5. Yi, S., Li, C., Li, Q.: A survey of fog computing: concepts, applications and issues. In: Proceedings of the 2015 Workshop on Mobile Big Data, pp. 37–42, June 2015
6. Costa, B., Bachiega Jr, J., Carvalho, L.R., Rosa, M., Araujo, A.: Monitoring fog computing: a review, taxonomy and open challenges. Comput. Netw. **215**, 109189 (2022)
7. Topcuoglu, H., Hariri, S., Wu, M.Y.: Performance-effective and low-complexity task scheduling for heterogeneous computing. IEEE Trans. Parallel Distrib. Syst. **13**(3), 260–274 (2002)
8. Aladwani, T.: Scheduling IoT healthcare tasks in fog computing based on their importance. Procedia Comput. Sci. **163**, 560–569 (2019)
9. Rehman, S., Javaid, N., Rasheed, S., Hassan, K., Zafar, F., Naeem, M.: Min-min scheduling algorithm for efficient resource distribution using cloud and fog in smart buildings. In: Barolli, L., Leu, FY., Enokido, T., Chen, HC. (eds.) Advances on Broadband and Wireless Computing, Communication and Applications. BWCCA 2018. LNDECT, vol. 25, pp. 15–27. Springer, Cham (2019). https://doi.org/10.1007/978-3-030-02613-4_2
10. Choudhari, T., Moh, M., Moh, T.S.: Prioritized task scheduling in fog computing. In Proceedings of the ACMSE 2018 Conference, pp. 1–8, March 2018
11. Bitam, S., Zeadally, S., Mellouk, A.: Fog computing job scheduling optimization based on bees swarm. Enterp. Inform. Syst. **12**(4), 373–397 (2018)
12. Aburukba, R.O., AliKarrar, M., Landolsi, T., El-Fakih, K.: Scheduling Internet of things requests to minimize latency in hybrid fog-cloud computing. Futur. Gener. Comput. Syst. **111**, 539–551 (2020)
13. Subbaraj, S., Thiyagarajan, R.: Performance oriented task-resource mapping and scheduling in fog computing environment. Cogn. Syst. Res. **70**, 40–50 (2021)
14. Vispute, S.D., Vashisht, P.: Energy-efficient task scheduling in fog computing based on particle swarm optimization. SN Comput. Sci. **4**(4), 391 (2023)
15. Saif, F.A., Latip, R., Hanapi, Z.M., Shafinah, K.: Multi-objective grey wolf optimizer algorithm for task scheduling in cloud-fog computing. IEEE Access **11**, 20635–20646 (2023)
16. Yoon, K.P., Hwang, C.L.: Multiple Attribute Decision Making: an Introduction. Sage Publications, Thousand Oaks (1995)

Dynamic Underload Host Detection for Performance Enhancement in Cloud Environment

Deepak Kumar Singh Yadav$^{(\boxtimes)}$ and Bharati Sinha

Computer Engineering Department, NIT Kurukshetra, Thanesar, India
{deepak_32123201,bharatisinha}@nitkkr.ac.in

Abstract. Cloud computing provides on-demand availability of computing resources, data storage and computing power. Cloud service providers often have functions distributed over multiple locations, each of which is a data center.Cloud computing relies on sharing of resources to achieve coherence and typically uses a pay-as-you-go model. However, cloud computing faces some significant challenges in efficiently managing resources, optimizing performance, and reducing energy consumption. Among the mentioned challenges ensuring optimal energy consumption is the key concern. Further, improvisation in energy efficiency helps minimize carbon emissions and also enhances overall performance. One of the primary reason of energy misuse in computation is host underload i.e., the host is not operating on its optimum capacity. The challenge of host underload detection, can be efficiently managed with the help of linear regression method. Our proposed approach aims to simultaneously reduce consumption of energy, minimize virtual machine (VM) migration and uphold SLA (Service Level Agreement) compliance. Any reduction in the number of VM migrations, results in better resource utilization and also mitigates the impact on performance caused by frequent migrations. This approach seeks to strike a balance among energy efficiency and meeting SLA, without compromising quality of service provided.

Keywords: Cloud Computing · Host Overload · Cloud datacenter · QoS · SLA

1 Introduction

CC has transformed the operational landscape for businesses by offering convenient and immediate access to a diverse array of computational resources for example servers, storage, and applications. This innovation empowers businesses to store and process data remotely, eliminating the requirement for extensive physical infrastructure and dedicated IT personnel [1]. CC offers numerous benefits, including scalability, cost-effectiveness and flexibility. Moreover, the exponential rise of CC has also brought about several challenges, one of which is host underload.

Host underload denotes to a situation in cloud computing where a physical server or virtual machine (VM) is operating with a workload significantly lower than its

K. K. Patel et al. (Eds.): icSoftComp 2023, CCIS 2031, pp. 29–40, 2024.
https://doi.org/10.1007/978-3-031-53728-8_3

capacity. It means that the resources allocated to the host are not fully utilized, resulting in inefficient resource allocation and potential wastage as show in Fig. 1. Underload hosts may consume unnecessary power, leading to increased energy costs and decreased operational efficiency [1, 2]. Addressing host underload involves strategies such as load balancing, dynamic resource allocation, and workload consolidation to optimize resource utilization, improve performance, and achieve cost-efficiency in cloud computing environments.

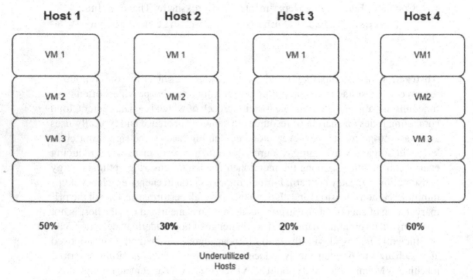

Fig. 1. Underutilized Host.

To solve this problem we proposed the algorithm uses linear regression technique for underload host detection in cloud computing. Underload host detection can significantly impact the SLA, Energy, and Performance. Linear regression is a widely utilized machine learning methodology i.e. commonly employed to forecast the value of a variable based on another variable [3]. Within this particular framework, the variable that is to be anticipated is commonly referred to as the dependent variable (y), whereas the variable employed for the purpose of prediction is denoted as the independent variable (x).

2 Literature Review

This section presents different strategies to overcome the underload host problem, reduce energy consumption and SLA violation and improve performance.

Youssef Saadi proposed an algorithm that focuses on energy efficiency for the consolidation of VM in cloud data centers. The aim is to minimize energy consumption while considering the utilization of hosts and ensuring the data center operates at optimal throughput. The authors compare their suggested strategy with onset algorithms, namely IQR and LR, and demonstrate its hopped-up [1]. The scheme successfully meets SLA requirements, as indicated by the simulation results.

Minarolli proposed a methodology for local resource allocation that involves modifying the CPU allocation assigned to virtual machines (VMs) in response to the current workload. Additionally, global resource allocation is achieved by VM migration, which aims to balance the distribution of resources among hosts that are either overloaded or underloaded. In order to forecast resource utilization, the authors utilize Gaussian processes as a machine learning methodology for time series prediction, taking into account long-term tendencies [2].

Abbas Horri, examines the trade-off that exists between energy usage and SLA Violation inside cloud systems. The objective of cloud service providers is to enhance their revenue by employing energy-efficient resource management strategies, which include the consolidation of virtual machines (VMs) and converting inactive servers into sleep modes. Nevertheless, inadequate consolidation may lead to the development of Sleep-Related Arousal Variants (SLAV). The author proposed the implementation of consolidation algorithms that aim to optimize energy usage while simultaneously minimizing Service Level Agreement Violations (SLAVs), in order to achieve a harmonious equilibrium between these two aims [3]. The simulation results indicate that the proposed methods effectively reduce the quantity of virtual machine migrations, SLAV (Service Level Agreement Violations), and total transferred data in comparison to existing strategies.

Nimisha introduced a novel algorithm, known as the Host Utilization Aware (HUA) Techniques, which aims to detect underloaded hosts. The algorithm presented in this study aims to estimate the upper limit of hosts that can be made available by taking into account the overall utilization of the data center. The primary emphasis of the Author's work is on the crucial stage of identifying underloaded hosts throughout the process of workload consolidation. The algorithm predicts the maximum number of hosts that may be made available by taking into account the overall utilization of the data center. The experimental results illustrate the effectiveness of the HUA Algorithm in accurately identifying hosts with low workload and thus freeing up a larger number of hosts. This leads to a reduction in energy consumption while still ensuring compliance with Service Level Agreement (SLA) requirements [4].

Weichao Ding revolves around the optimization of resource allocation and utilization in Cloud data centers. The overarching objective is to achieve a reduction in energy consumption while simultaneously upholding a high degree of compliance with service-level agreements (SLAs). The suggested framework by the author introduces a novel approach to dynamic virtual machine (VM) consolidation [5]. This approach is based on the prediction of resource use and the PPR (performance-to-power ratio) of heterogeneous hosts. The framework has four distinct stages, including host overload detection, VM selection for migration, host underload detection, and VM allocation with a modified power-aware best-fit reducing method. The proposed methodology effectively reconciles the trade-off between energy usage and performance. The suggested strategy has been validated for its effectiveness and scalability through experimental evaluations.

Bernardi Pranggono examines the matter of VM consolidation in cloud data centers and presents a novel approach for classifying host load inside an energy-performance VMC framework. The primary aim is to decrease energy usage while simultaneously guaranteeing the fulfillment of quality of service (QoS) criteria. The proposed approach entails the classification of hosts experiencing underload into three distinct states: underloaded, normal, and critical. This classification is achieved by the utilization of an underload detection algorithm. Additionally, the study presents the introduction of overload detection as well as the VM selection strategies. The overload detection policy, referred to as Mean (Mn), utilizes the mean to forecast the upper threshold [6]. On the other hand, the VM selection policy, known as Maximum Requested Bandwidth (MBW), relies on the maximum requested bandwidth.

Nirmal Kr. Biswas revolves around the resolution of two key issues in Smart Cities: the optimization of energy consumption as well as the mitigation of SLA violation. To tackle these challenges, Biswas proposes the utilization of Cloud of Things (CoT) technology. The growing processing capabilities inside cloud computing have necessitated the identification of an optimal balance between energy consumption and service level agreement violations (SLAV). The proposed methodology put forth by the author involves the integration of a novel New Linear Regression prediction model, host underload/overload detection as well as a VM placement policy. The objective of this technique is to effectively mitigate both energy consumption and service level agreement violations. The new linear regression prediction model has been developed with the purpose of forecasting forthcoming CPU use by employing a linear regression line and a mean point (MNP) [7]. The proposed algorithms are assessed using an expanded version of the CloudSim Simulator.

The suggested techniques are evaluated using CloudSim simulation with different types of random workloads as well as PlanetLab real. MadMCHD algorithm shows significant improvements compared to commonly used algorithms such as thr, mad, iqr, lr, and lrr, in minimizing SLA violation rates and VM migrations. The combination of MadMCHD and MPABFD algorithms further reduces SLA violations overall.

3   Proposed Work

In this section, we first briefly introduce Linear Regression, then show the detail of our suggested Linear Regression algorithm for underload host detection. With the help of this algorithm, we decrease energy consumption and improve the performance matrix.

3.1   Linear Regression

Linear regression is a fundamental technique in the field of machine learning, commonly employed for the purpose of predicting the value of a variable based on the value of another variable. The variable to be predicted is commonly referred to as the dependent variable (y). The variable employed for predicting the value of other variables is commonly referred to as the independent variable (x). The objective of the linear regression algorithm is to determine the optimal values for B0 and B1 in order to identify the most suitable line of best fit. The best fit line is characterized by its ability to minimize the error between projected values and actual values.

To calculate linear regression best fit line.

$$Yi = \beta0 + \beta1Xi \tag{1}$$

where Xi = Independent variable, Yi = Dependent variable, $\beta1$ = Slope/Intercept, $\beta0$ = constant/Intercept (Fig. 2).

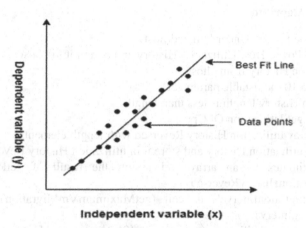

Fig. 2. Linear Regression.

Here's a step-by-step explanation of how to use linear regression for dynamic underload prediction.

- Data Collection
- Data Preprocessing
- Split the Data
- Model Building
- Model Training
- Prediction
- Monitoring and Updating

3.2 Energy Efficient Linear Regression Algorithm for Underload Host Detection

The proposed Energy efficient linear regression algorithm for the host underload detection can be formulated. In the proposed algorithm, we predict the upcoming load on the host on behave of the previous host load history and set the minimum threshold value. The host is considered underutilized whenever the host load value is down from the minimum threshold value [9]. This proposed algorithm is helpful in improving energy efficiency and also improving performance.

3.3 Proposed Algorithm

1. Begin method is Host Under Utilized(host)
2. Cast host to Power Host Utilization History and assign it to _host
3. Get utilization History from _host
4. Set length to 10 (adjustable parameter)
5. If utilization History length is less than length,
6. Return host.get Utilization Of Cpu()
7. Create an array utilization History Reversed with length elements
8. Reverse the utilization History and store it in utilization History Reversed
9. Declare estimates as an array and assign the result of getParameterEstimates(utilizationHistoryReversed)
10. Calculate migrationIntervals as ceil (getMaximumVmMigrationTime(_host) / getSchedulingInterval()).
11. Calculate predictedUtilization as (estimates[0] + estimates[1] * (length + migrationIntervals)) * getSafetyParameter()
12. Add a history entry with host and predictedUtilization
13. Return predictedUtilization
14. End method is HostUnderUtilized.

In this algorithm, we take the three-parameter for enhancing performance. The First parameter is SLA violation second parameter is Energy, and the last parameter is No of VM migration.

$$PM = E * SLA * No\ Of\ VM\ Migration \qquad (2)$$

where PM is Performance matrix, E is Energy, SLA is Service level agreement.

This formula is used for calculating overall performance, and after calculating overall performance, we quickly identify which technique gives the best result (Fig. 3).

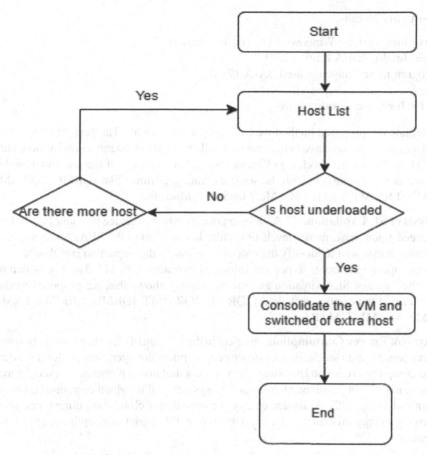

Fig. 3. Workflow Diagram of the Proposed Algorithm.

4 Result

4.1 Simulation Setup

The presented technique is implemented using the CloudSim toolkit 3.0.3 simulator. CloudSim is specifically designed to simulate various components of a cloud system, including virtual machines (VMs) and data centers. It offers support for VM selection and allocation policies, power models, and diverse workload types. The hardware and software requirements for implementing the proposed method are outlined as follows.

Hardware Details. The hardware used in the implementation of the proposed method.

- Processor: Intel(R) Core(TM) i3-6100U CPU @ 2.30 GHz
- RAM: 8 G.B
- Hard disk: 2 TB

Software Details.

- Operating system: Windows 10 Pro 64-bit Version
- IDE: IntelliJ IDEA Edition 2021.3
- Programming Language used: JAVA 17.0.1

4.2 Performance Evaluations

To simulate the proposed method we have used the cloudsim. The performance of proposed technique is measured in the term of the SLA violation, energy consumption, number VM migration. To check the effectiveness and correctness of the presented method we have done comparison with the some existing algorithm like IQRMC, IQRMMT, IQRMU, LRMMT, LRMU, MadMc, MadMMT, MadMu.

Analysis of SLA Voilation. SLA violation occurs when a service provider fails to meet the agreed-upon performance levels or standards specified in the SLA. Analyzing SLA violations is essential to identify the root causes, assess the impact on stakeholders, and take appropriate measures to prevent future occurrences [10, 11]. The Fig. 4 demonstrate the value of SLA violation and finding clearly shows that our proposed method outperformed the existing techniques IQRMC, IQRMMT, IQRMU, LRMMT, LRMU, MadMc, MadMMT, MadMu.

Analysis of Energy Consumption. It refers to the amount of electrical power consumed by data centers and associated infrastructure to support the operation of cloud services. Cloud computing relies on large-scale data centers that house numerous servers, storage systems, networking equipment, and cooling systems, all of which consume significant amounts of energy [12]. Analyzing energy consumption in cloud computing is crucial for improving energy efficiency, reducing environmental impact, and optimizing resource utilization.

Energy consumption of data center is determine with the given formula.

$$(EC) = E_1 + E_1 + -E_n \tag{3}$$

where E_i = energy consumption of host. E_i Can be calculate using linear interpolation method based on given utilization.

Figure 5 demonstrate the results of the energy consumption of presented approach compare with the other existing techniques and results clearly indicates that presented technique perform better in terms of energy saving.

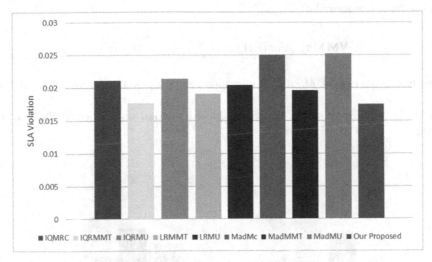

Fig. 4. Analysis SLA Violation.

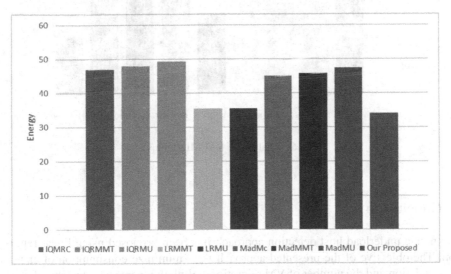

Fig. 5. Analysis of Energy Consumption.

Analysis of VM Migration. VM migration is a process of moving a running or idle virtual machine instance from one physical host or data center to another. This is done for various reasons, including load balancing, resource optimization, hardware maintenance, and disaster recovery. The number of VM metrics can be calculated by utilizing the below formula:

$$\text{VM Mirgration}(F,t1,t2) = \sum_{i=1}^{N} \int_{t1}^{t2} Mig_i(F) \tag{3}$$

where, N is count of VM, $Mig_i(F)$ is migration count of host i in the time interval t1 to t2.

The Fig. 6, clearly shows that the proposed method outperformed the existing techniques.

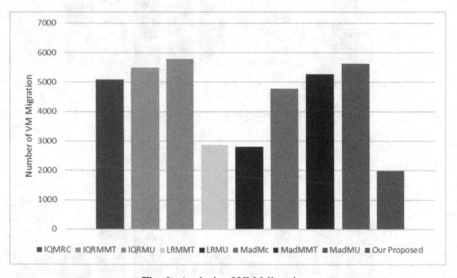

Fig. 6. Analysis of VM Migrations.

5   Conclusion

A dynamic underload host detection approach has been proposed using linear regression. The objective of the presented approach is to minimize consumption of energy, SLA violation and the number of VM migrations. With the increasing demand of cloud computing the energy consumption is also rising sharply. Hence, energy saving becomes prime concern in cloud because it also helps in reduction of cost, reduction in carbon emission and it also increase the performance of the cloud. The efficiency and accuracy of the presented technique is evaluated in terms of SLA violations, consumption of energy, and the number of migrations as performance metrics. The experiments are conducted over cloudsim tool. The performance of the presented technique is compared with the standard existing techniques like IQMRC, IQRMMT, IQRMU, LRMMT, LRMU, MADMC, MADMMT, MADMU. The results shows that the presented approach outperforms existing techniques in terms of consumption of energy, SLA violation as well as number of VM migrations. In the future, we can leverage real-time data for host underload prediction to improve the accuracy and efficacy of the proposed algorithm in real-time scenarios.

References

1. Saadi, Y., El Kafhali, S.: Energy-efficient strategy for virtual machine consolidation in cloud environment. Soft. Comput. **24**(19), 14845–14859 (2020)
2. Minarolli, D., Mazrekaj, A., Freisleben, B.: Tackling uncertainty in long term predictions for host overload and underload detection in cloud computing. J. Cloud Comput. **6**, 1–18 (2017)
3. Horri, A., Mozafari, M.S., Dastghaibyfard, G.: Novel resource allocation algorithms to performance and energy efficiency in cloud computing. J. Supercomput. **69**, 1445–1461 (2014)
4. Patel, N., Patel, H.: Energy efficient strategy for placement of virtual machines selected from underloaded servers in compute cloud. J. King Saud Univ.-Comput. Inf. Sci. **32**(6), 700–708 (2020)
5. Ding, W., Luo, F., Han, L., Gu, C., Lu, H., Fuentes, J.: Adaptive virtual machine consolidation framework based on performance-to-power ratio in cloud data centers. Futur. Gener. Comput. Syst. **111**, 254–270 (2020)
6. Alboaneen, D.A., Pranggono, B., Tianfield, H.: Energy-aware virtual machine consolidation for cloud data centers. In: 2014 IEEE/ACM 7th International Conference on Utility and Cloud Computing, pp. 1010–1015. IEEE, December 2014
7. Biswas, N.K., Banerjee, S., Biswas, U., Ghosh, U.: An approach towards development of new linear regression prediction model for reduced energy consumption and SLA violation in the domain of green cloud computing. Sustain. Energy Technol. Assess. **45**, 101087 (2021)
8. Hsieh, S.Y., Liu, C.S., Buyya, R., Zomaya, A.Y.: Utilization-prediction-aware virtual machine consolidation approach for energy-efficient cloud data centers. J. Parallel Distrib. Comput. **139**, 99–109 (2020)
9. Li, L., Dong, J., Zuo, D., Wu, J.: SLA-aware and energy-efficient VM consolidation in cloud data centers using robust linear regression prediction model. IEEE Access **7**, 9490–9500 (2019)
10. Wang, J., Gu, H., Yu, J., Song, Y., He, X., Song, Y.: Research on virtual machine consolidation strategy based on combined prediction and energy-aware in cloud computing platform. J. Cloud Comput. **11**(1), 1–18 (2022)
11. Melhem, S.B., Agarwal, A., Goel, N., Zaman, M.: Markov prediction model for host load detection and VM placement in live migration. IEEE Access **6**, 7190–7205 (2017)
12. Kulshrestha, S., Patel, S.: An efficient host overload detection algorithm for cloud data center based on exponential weighted moving average. Int. J. Commun. Syst. **34**(4) (2021)
13. Daraghmeh, M., Melhem, S.B., Agarwal, A., Goel, N., Zaman, M.: Linear and logistic regression based monitoring for resource management in cloud networks. In: 2018 IEEE 6th International Conference on Future Internet of things and Cloud (FiCloud), pp. 259–266. IEEE, August 2018
14. Abdelsamea, A., El-Moursy, A.A., Hemayed, E.E., Eldeeb, H.: Virtual machine consolidation enhancement using hybrid regression algorithms. Egyptian Inf. J. **18**(3), 161–170 (2017)
15. Alhammadi, A.S.A., Vasanthi, V.: Multiple regression particle swarm optimization for host overload and under-load detection. TEST Eng. Manag. **17**(2), 1109 (2020)

16. Nehra, P., Nagaraju, A.: Host utilization prediction using hybrid kernel based support vector regression in cloud data centers. J. King Saud Univ.-Comput. Inf. Sci. **34**(8), 6481–6490 (2022)
17. A El-Moursy, A., Abdelsamea, A., Kamran, R., Saad, M.: Multi-dimensional regression host utilization algorithm (MDRHU) for host overload detection in cloud computing. J. Cloud Comput. **8**(1), 1–17 (2019)
18. Jararweh, Y., Issa, M.B., Daraghmeh, M., Al-Ayyoub, M., Alsmirat, M.A.: Energy efficient dynamic resource management in cloud computing based on logistic regression model and median absolute deviation. Sustain. Comput. Inf. Syst. **19**, 262–274 (2018)
19. Yadav, R., Zhang, W., Li, K., Liu, C., Shafiq, M., Karn, N.K.: An adaptive heuristic for managing energy consumption and overloaded hosts in a cloud data center. Wireless Netw. **26**(3), 1905–1919 (2020)
20. Sissodia, R., Rauthan, M.S., Barthwal, V.: A multi-objective adaptive upper threshold approach for overloaded host detection in cloud computing. Int. J. Cloud Appl. Comput. (IJCAC) **12**(1), 1–14 (2022)
21. Hema, M., Raja, S.: An efficient framework for utilizing underloaded servers in compute cloud. Comput. Syst. Sci. Eng. **44**(1), 143–156 (2023)
22. Mao, L., Chen, R., Cheng, H., Lin, W., Liu, B., Wang, J.Z.: A resource scheduling method for cloud data centers based on thermal management. J. Cloud Comput. **12**(1), 1–18 (2023)

Inverse Reinforcement Learning to Enhance Physical Layer Security in 6G RIS-Assisted Connected Cars

Sagar Kavaiya[1]([✉]), Narendrakumar Chauhan[2], Purvang Dalal[2],
Mohitsinh Parmar[1], Ravi Patel[1], and Sanket Patel[1]

[1] Smt. Chandaben Mohanbhai Patel Institute of Computer Applications, Charotar
Univesity of Science and Technology, Changa, India
{sagarkavaiya.mca,mohitsinhparmar.mca,ravipatel.mca,
sanketpatel.mca}@charusat.ac.in
[2] Faculty of Technology, Dharmsinh Desai University, Nadiad, India
{nvc.ec,pur_dalal.ec}@ddu.ac.in

Abstract. This research paper introduces a groundbreaking approach
to address the escalating security concerns in the era of 6G communication networks, particularly in the context of Reconfigurable Intelligent
Surfaces (RIS)-assisted connected cars. With the proliferation of connected vehicles, ensuring the confidentiality of wireless communications
has become paramount. In response, this study harnesses the potential of
Inverse Reinforcement Learning (IRL) to fortify the physical layer security of such networks. By integrating IRL into the RIS-assisted vehicular
communication framework, a novel strategy emerges. Firstly, the paper
formulates the eavesdropper's potential actions as a learning problem,
allowing the system to discern and adapt to potential threat scenarios.
Leveraging this acquired knowledge, the RIS optimizes the configuration of its reflective elements dynamically, thwarting the eavesdropper's
attempts and bolstering communication security. Secondly, the extensive
derivations of performance metrics - signal-to-interference noise ratio and
bit error rate have been carried out to show the importance of IRL approach. Through comprehensive simulations, the efficacy of the proposed
IRL-infused mechanism is validated, demonstrating significant advancements in communication privacy compared to conventional methods.
This research bridges the domains of 6G networks, vehicular technology, and machine learning, presenting a promising avenue to reinforce
the safeguarding of connected cars against emerging security challenges.
As the 6G landscape unfolds, the fusion of innovative techniques like IRL
holds immense potential to reshape security paradigms in vehicular communications, amplifying the resilience of next-generation transportation
systems.

Keywords: 6G · Physical Layer Security · Inverse Reinforcement
Learning

K. K. Patel et al. (Eds.): icSoftComp 2023, CCIS 2031, pp. 41–53, 2024.
https://doi.org/10.1007/978-3-031-53728-8_4

1 Introduction

1.1 Background

Learning from demonstration, known as imitation learning, involves acquiring the ability to perform actions within an environment by observing examples presented by an instructor. Inverse Reinforcement Learning (IRL) is a specialized approach within imitation learning that strives to deduce the reward function of a Markov decision process using examples given by the instructor. The reward function essentially encapsulates the core essence of a task. While straightforward tasks might have a readily available reward function, which can be directly included in the learning process, complex tasks may lack this straightforwardness. In such scenarios, it becomes more feasible to learn the reward function by analyzing the actions undertaken by the instructor. Autonomous vehicles offer a dual advantage: enhancing road safety and concurrently boosting fuel efficiency while alleviating traffic congestion. They constitute the predominant trajectory within forthcoming intelligent transportation systems [1].

IRL holds significance within this context due to its role in teaching autonomous vehicles to operate effectively. IRL assists in determining the underlying reward function that governs decision-making in complex environments. For autonomous vehicles, this translates to the capability to understand human driving behaviors and intentions, enabling them to mimic human-like actions in diverse scenarios. By discerning the reward structure through IRL, autonomous vehicles can better comprehend optimal behaviors, making them safer, more efficient, and seamlessly integrated into existing traffic systems [2].

IRL is vital for bolstering physical layer security (PLS) by deciphering adversary actions in wireless systems. It models adversarial behaviors, informs countermeasure design, adapts to evolving threats, and optimizes resource allocation. IRL minimizes false positives, integrates human behavior, and enhances security by learning reward structures, fortifying wireless networks against emerging risks while ensuring efficient communication [3–5].

1.2 Related Works

This paper [2] focuses on planning for autonomous vehicles in traffic to enhance safety, fuel efficiency, and congestion reduction. It employs reinforcement learning and inverse reinforcement learning, using a stochastic Markov decision process. Expert driving behavior is learned from demonstrations, and a deep neural network approximates the reward function. Simulated results validate the approach. In addition to that the work in [1] offers a thorough overview of learning from demonstration, specifically focusing on IRL. IRL aims to deduce the reward function in complex tasks by observing a teacher's actions. The survey distinguishes IRL from related methods, outlines its applications, and suggests avenues for future research in this field. Furthermore, this work [6] introduces an online data-driven model-based inverse reinforcement learning (MBIRL) technique for both linear and nonlinear deterministic systems. It estimates unknown reward and optimal value functions in real-time from observed agent trajectories, using a novel feedback-driven approach. Theoretical guarantees for error convergence are provided, and numerical experiments validate its efficacy in solving inverse reinforcement learning problems.

The aforementioned work mainly tackles the learning problem to plan resource utilisation effectively. However, several works have been reported into literature that tackles the secrecy problem of wireless channel with reinforcement learning (RL). The work in [7] addresses the optimization of 3D trajectories for Unmanned Aerial Vehicles (UAVs) assisted by Reconfigurable Intelligent Surfaces (RIS) to maximize physical layer security rates during wireless transmission. Traditional optimization methods struggle due to non-convexity, so the paper introduces a Double Deep Q Network (DDQN)-based reinforcement learning approach. Simulation results demonstrate its superiority over other methods, enhancing safety rates significantly. To implement the IRL with PLS, this paper [4] explores the influence of Deep Learning on wireless network security, particularly in countering intelligent attackers who can adapt to standard defense mechanisms. It introduces two intelligent defense mechanisms based on inverse reinforcement learning, successfully enhancing defense capabilities. Experimental tests against intelligent attackers in a backoff attack scenario yield substantial performance improvements.

Handling the 6G networks, the work in [8] addresses the security and privacy challenges inherent in 6G cellular systems, characterized by heterogeneous networks and advanced technologies. It emphasizes the limitations of existing optimization-based security approaches and advocates for Reinforcement Learning (RL) algorithms to adapt to dynamic network conditions and counteract smart attacks. The survey covers potential attacks, RL-based security solutions, and identifies future research directions for 6G systems.

1.3   Motivations and Novelties

The existing research papers mentioned in the above section mainly focus on various aspects of reinforcement learning, inverse reinforcement learning, and their applications in optimizing different domains. While they cover areas such as autonomous vehicle planning, model-based inverse reinforcement learning, and security in wireless networks, they primarily address resource allocation, learning from demonstrations, and defense mechanisms against attacks. However, there is a noticeable gap in addressing the specific challenges of physical layer security in 6G RIS-assisted connected cars.

Firstly, the application domain of connected cars, particularly those integrating 6G technology and Reconfigurable Intelligent Surfaces (RIS), represents a unique and rapidly evolving landscape. The need to establish robust physical layer security within this context is paramount, yet there is a noticeable absence of dedicated research in this specialized area.

Secondly, the emergence of 6G networks introduces a fresh set of security challenges characterized by high data rates, minimal latency, and extensive connectivity. Effectively addressing these challenges is pivotal for the successful integration and deployment of 6G-connected cars, which are expected to play a central role in the future of transportation.

Moreover, connected cars routinely handle and transmit sensitive data, which underscores the critical importance of safeguarding user privacy and ensuring vehicle safety. Any potential vulnerability in the physical layer security could have far-reaching consequences, further emphasizing the urgency of developing advanced security mechanisms tailored to this unique environment.

Since RIS are identified as a key enabler of 6G networks, this offers the potential to enhance communication performance. Leveraging RIS for physical layer security in connected cars presents a promising avenue, but it necessitates specialized security solutions, making it an intriguing area for exploration.

Additionally, the evolving landscape of cyber threats poses a significant challenge. These threats are increasingly sophisticated and adaptive, making traditional security measures less effective. Thus, there is a clear need for intelligent and adaptable security mechanisms, and reinforcement learning techniques are well-suited to address this requirement.

Furthermore, the research acknowledges the interdisciplinary nature of addressing physical layer security in connected cars, requiring expertise from diverse fields such as wireless communication, machine learning, security, and automotive engineering. This interdisciplinary approach is essential to formulate comprehensive security solutions. Given the imminent deployment of 6G-connected cars in real-world scenarios, the research directly tackles a pressing and practical issue. Enhancing physical layer security in these vehicles is not only essential but also timely, as it will significantly impact the safety and security of future transportation systems.

1.4   Contributions

The main contributions of this paper can be summarised as follows:

1. Development of an Innovative Security Framework: The paper introduces an innovative security framework that leverages Reflective Intelligent Surfaces (RIS) in the context of 6G networks. It outlines a novel approach to enhancing physical layer security in connected cars, addressing the unique challenges posed by this emerging technology.
2. Integration of Inverse Reinforcement Learning (IRL): One of the key contributions is the incorporation of IRL as a central component of the security framework. This novel application of IRL allows the system to autonomously adapt and optimize communication strategies to thwart eavesdropping attempts and malicious attacks, thereby improving the overall security posture.
3. Dynamic Threat Assessment and Mitigation: The research paper presents a dynamic threat assessment mechanism that continuously evaluates the security landscape in real-time. By using IRL, the system can adapt its security measures on-the-fly, making it highly responsive to evolving threats and vulnerabilities.

4. Optimization of RIS Parameters: The paper provides insights into how RIS parameters, such as the phase shift and reflection coefficients, can be optimized to enhance security without compromising communication quality. This optimization process is driven by the IRL algorithm, which learns from historical data and adapts to changing conditions.

5. Performance Evaluation and Validation: The research conducts extensive simulations and real-world experiments to validate the effectiveness of the proposed security framework. It quantitatively demonstrates the improvements in physical layer security achieved through the integration of IRL and RIS in connected car scenarios.

2 System Model

2.1 6G Channel Model

The Wyner's wiretap channel model is considered to obtain the performance over fading model [9]. The fading is modeled by considering the dynamic 2×2 Nakagami-m channel between the legitimate transmitter, legitimate receiver and eavesdropper. The channel state information is considered to be perfect. Based on the joint eigenvalues of 2×2 Nakagami-m, the distribution of received SNR can be expressed as follows [10, Eq. 23]:

$$p(L) = \frac{K_{22}\, 8\, \pi^{5/2}}{2^{2m-1}} \times \frac{m \mathrm{tr}(LL')}{\Omega} l_{11}^{4m-1}$$

$$\sum_{i_1=0}^{m-1} \sum_{k_1=0}^{i_1} \sum_{i_2=0}^{m-1} \sum_{k_2=0}^{i_2} \binom{m-1}{i_1}\binom{i_1}{k_1}\binom{m-1}{i_2}\binom{i_2}{k_2} \times t_1 \times t_2 \quad (1)$$

where,

$$t_1 = (-1)^{m-1-i_2}(l_{21I}^2 + l_{21R}^2)^{k_1+k_2+1/2(2m-i_1-i_2-2)}$$

$$l_{22}^{i_2-2k_2+i_1-2k_1+2m-1}\Gamma(1/2(2m - i_1)) \quad (2)$$

$$t_2 = \frac{2^{4m-4-i_1-i_2}\Gamma(1/2(4m + 2k_1 + i_2 - 2k_2 - 2 - i_1))}{\Gamma(4(m-1))}$$

$$\frac{\Gamma(1/2(i_1 - 2k_1 + 2k_2 + 4m))}{\Gamma(1/2(2m - i_1 - i_2))} \quad (3)$$

m is the fading parameter. $\Gamma(\cdot)$ is the Gamma function. $K_{ij} = \left(\frac{m^m}{\pi \Omega^m \Gamma(m)}\right)^{ij}$, it can be noticed that at $m = 1$, P(L) reduced to the uniform distribution on the circle of radius Ω. Furthermore $f(i_1, i_2, k_1, k_2)$ can be considered as [10, Eq. 14], L is the number of branches, l is the multi-path component between the transmitter and receiver antenna.

2.2 6G Signal Model

Combining the subset of receive antennas with the largest SNRs at legitimate receiver results in the instantaneous SNR in the main channel as

$$\gamma_R = \sum_{l_R=1}^{L_R} \gamma_{(l_R)}^R \tag{4}$$

The above expression implies the received signal at legitimate receiver side. Furthermore, In the eavesdropper's channel, at time slot l, the received signal vector is given by

$$\mathbf{y}_E(l) = \mathbf{h}x(l) + \mathbf{n}_E(l) \tag{5}$$

2.3 Mobility Model

The $\mathbf{h}$ is further expressed as a function of distance between legitimate transmitter and eavesdropper as follows [11]:

$$h_{(t)} = \frac{g_x}{\sqrt{1 + d_X^\alpha(t)}}, \tag{6}$$

where α denotes the path loss exponent, g_x is the channel gain over MIMO Nakagami-m distribution. d_X is the distribution of distance.

3 Derivation of Signal-to-Interference-Plus-Noise Ratio (SINR)

The SINR is a crucial performance metric in wireless communication systems, including those involving Reflective Intelligent Surfaces (RIS). It quantifies the quality of the received signal relative to interference and noise. The SINR can be derived as follows: Consider a scenario with a transmitter, a receiver, and a reflective intelligent surface (RIS) in the channel. Let P_t be the transmitted power, h_d be the channel gain between the transmitter and receiver, h_r be the channel gain between the RIS and the receiver, and N be the noise power. The received signal power, P_{signal}, is given by:

$$P_{\text{signal}} = |h_d|^2 P_t$$

The interference power, $P_{\text{interference}}$, from other sources (e.g., adjacent users or noise from RIS reflections) is given by:

$$P_{\text{interference}} = |h_r|^2 P_t$$

The noise power, P_{noise}, is constant and equal to N. Therefore, the SINR is calculated as:

$$\text{SINR} = \frac{P_{\text{interference}} + P_{\text{noise}}}{P_{\text{signal}}} = \frac{|h_r|^2 P_t + N}{|h_d|^2 P_t}$$

4   Derivation of Bit Error Rate (BER)

The Bit Error Rate (BER) measures the probability of incorrect bit reception due to noise and interference. It's a key performance metric in assessing the reliability of data transmission. The BER can be derived as follows: Let S be the received signal amplitude for a transmitted bit '1', and N_0 be the one-sided power spectral density of the noise. The received signal for bit '0', denoted as S_0, is assumed to be zero for simplicity (ideal signaling):

$$S_0 = 0$$

The decision threshold is typically set at $0.5(S + S_0)$. The BER for a binary modulation scheme (e.g., binary phase-shift keying, BPSK) can be expressed as:

$$\text{BER} = \frac{1}{2}\text{erfc}\left(\sqrt{\frac{2N_0}{S}}\right)$$

where erfc is the complementary error function. This equation quantifies the probability of bit errors as a function of signal power, noise power, and modulation scheme. It allows you to assess the system's performance under varying conditions. These derivations provide mathematical expressions for key performance metrics that can be used to evaluate and optimize the proposed security framework for 6G RIS-assisted connected cars in terms of signal quality and data transmission reliability. Further derivations may be needed to model other performance metrics or specific aspects of the research.

5   Improving Secrecy Using IRL

5.1   State and Action Spaces

Define the state space $\mathcal{S}$, which includes all relevant parameters that describe the current state of the system. For example, $\mathcal{S}$ may encompass channel conditions (Channel), Reflective Intelligent Surface (RIS) configuration (RIS), and positions of vehicles (Vehicle). Similarly, define the action space $\mathcal{A}$, representing actions that can optimize security. Actions may involve adjusting RIS phase shifts (θ), transmission power (P), or other parameters.

5.2   Reward Function

Design a reward function $R(s, a)$ that quantifies security and communication quality. This function should consider factors like received signal power (P_{signal}), interference ($P_{\text{interference}}$), eavesdropping probability ($P_{\text{eavesdropping}}$), data rate (R_{data}), among others. The goal is to maximize this reward function while ensuring secure communication.

$$R(s, a) = f(P_{\text{signal}}, P_{\text{interference}}, P_{\text{eavesdropping}}, R_{\text{data}}, \ldots) \tag{7}$$

5.3   Transition Dynamics

Model how the system transitions from one state to another based on selected actions. The transition function $\mathcal{T}$ captures how changes in RIS configuration, transmission power, or other actions impact the system's state.

$$s_{t+1} = \mathcal{T}(s_t, a_t) \tag{8}$$

5.4   Inverse Reinforcement Learning (IRL)

IRL is employed to learn the reward function from observed data. Collect data from agent-environment interactions, where the connected car takes actions to optimize security. The observed data is then used to infer the underlying reward function that the agent seeks to maximize.

5.5   Optimization Objective

Once the reward function is learned through IRL, formulate an optimization problem to find the optimal actions that maximize the learned reward function while adhering to constraints. This can be formulated as a constrained optimization problem:

$$\text{Maximize } R(s, a)$$
$$\text{Subject to } a \in \mathcal{A} \tag{9}$$

6   Solution Method

To solve the optimization problem maximizing the reward function $R(s, a)$, we employ gradient-based methods. Let $\nabla R(s, a)$ denote the gradient of the reward function with respect to the action a. Our objective is to find the optimal action a^* that maximizes the reward:

$$a^* = \arg\max_a R(s, a). \tag{10}$$

We start with an initial action a_0 and iteratively update it to find the optimal action. The update rule is given by:

$$a_{t+1} = a_t + \alpha \nabla R(s, a_t), \tag{11}$$

where α is the learning rate that controls the step size in each iteration.

6.1   Gradient Ascent

In gradient ascent, we aim to maximize the reward by iteratively moving in the direction of the gradient. The update rule for gradient ascent is:

$$a_{t+1} = a_t + \alpha \nabla R(s, a_t). \tag{12}$$

Algorithm 1. Enhancing Physical Layer Security for Connected Cars using IRL

1: **Step 1:** Data Collection
2: $D \leftarrow$ collect_communication_data()
3: **Step 2:** Feature Extraction
4: $F \leftarrow$ extract_features(D)
5: **Step 3:** Reward Function Inference using IRL
6: $S, A, \mathcal{D} \leftarrow$ preprocess_data(F)
7: $R \leftarrow$ infer_reward_function($S, A, \mathcal{D}$)
8: **Step 4:** Intent Inference
9: $\mathcal{M} \leftarrow$ interpret_reward_function(R)
10: $\mathcal{A} \leftarrow$ detect_attack_patterns($\mathcal{M}$)
11: **Step 5:** Security Strategy Optimization
12: $\mathcal{S} \leftarrow$ design_security_strategies(R)
13: $\mathcal{S}^{\leftarrow}$ optimize_strategies($\mathcal{S}$)
14: **Step 6:** Adaptive Defense
15: $\mathcal{M}_{\mathrm{dyn}} \leftarrow$ implement_dynamic_monitoring(D)
16: $\mathcal{S}_{\mathrm{adapt}} \leftarrow$ adapt_security_strategies($\mathcal{M}_{\mathrm{dyn}}, \mathcal{S}$)
17: **Step 7:** Evaluation and Validation
18: $\mathcal{P} \leftarrow$ simulate_performance($\mathcal{S}_{\mathrm{adapt}}, D$)
19: **Step 8:** Comparison and Analysis
20: $\mathcal{B} \leftarrow$ compare_baseline_performance(D)
21: analyze_results($\mathcal{P}, \mathcal{B}$)
22: **Step 9:** Implementation and Deployment
23: deploy_algorithm($\mathcal{S}_{\mathrm{adapt}}$)
24: **Step 10:** Fine-Tuning and Updates
25: continuous_update($\mathcal{S}_{\mathrm{adapt}}$, new_data)

We continue this process until convergence to find the optimal action a^* that maximizes the reward.

The proposed algorithm aims to enhance physical layer security in the context of connected cars using Inverse Reinforcement Learning (IRL). The algorithm addresses the pressing need for robust security measures in vehicular communication systems, where adversaries may exploit vulnerabilities to compromise safety and privacy. In the algorithm, the process begins with the collection of real-world communication data between connected vehicles and Reconfigurable Intelligent Surfaces (RIS). Relevant features are then extracted from this data to form a representation of communication behaviors and system dynamics.

The core of the algorithm lies in Step 3, where IRL is employed to infer a reward function that characterizes observed communication behaviors. This reward function encapsulates the underlying motivations and intentions of both legitimate actors and potential adversaries. By interpreting this reward function, the algorithm discerns the motivations behind adversarial actions, identifying patterns that indicate possible attacks.

Based on the inferred reward function, security strategies are designed and optimized in Step 5. These strategies dictate resource allocation, transmission power, and other parameters to counteract potential attacks while maintaining efficient communication. The algorithm is designed for adaptability and resilience. Step 6 introduces a dynamic defense mechanism that continuously monitors the communication environment, ensuring security strategies adapt to changing conditions. Performance evaluation in Step 7 and comparison with baseline methods in Step 8 provide quantitative insights into the algorithm's effectiveness.

The algorithm culminates in Steps 9 and 10, where the refined security strategies are deployed in real-world connected car systems. Regular updates and fine-tuning ensure the algorithm remains effective against emerging threats.

By leveraging Inverse Reinforcement Learning, this algorithm offers a comprehensive framework for strengthening the security of connected cars. It underscores the potential of machine learning techniques to enhance vehicular communication's resilience against adversarial actions and lays the foundation for safer and more secure connected transportation systems.

7   Numerical Results and Discussions

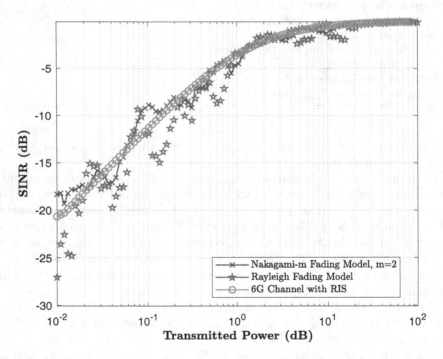

Fig. 1. SINR against Transmit Power for Various Channel Model (Noise power = 1, Transmit Power Range = 0.01 dB to 100 dB, m = 2, Window size = 5).

The result demonstrates the behaviour of the SINR for each channel model, Fig. 1 calculates SINR values for each channel model and each transmitted power value within the specified range. For the Nakagami-m channel, it generates channel gains using a gamma distribution and calculates SINR based on these gains. For the Rayleigh channel, it generates channel gains using exponential distribution and calculates SINR based on these gains. For the 6G channel with RIS, it uses provided values for channel gains. It helps in understanding how different channel models affect the SINR in a wireless communication system and provides insights into their relative performance under varying power conditions.

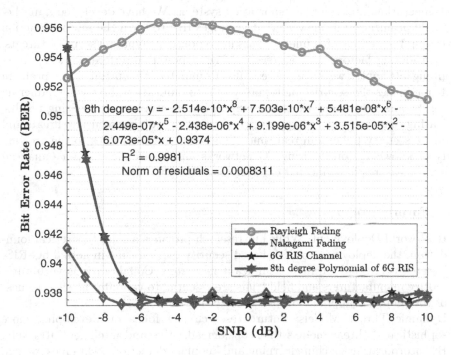

Fig. 2. BER against SNR for Various Channel Model (Noise power $= 1$, Transmit Power Range $= 0.01$ dB to 100 dB, m $= 2$, Window size $= 5$).

Figure 2 enables a direct comparison of the error performance of three distinct channel models. This comparison is crucial for understanding how different channel conditions impact the reliability of data transmission in wireless communication systems. This considers both Rayleigh and Nakagami fading models, allowing you to assess the influence of fading phenomena on BER. Rayleigh fading represents a multipath propagation scenario, while Nakagami fading introduces controlled variability based on the Nakagami-m parameter. Understanding these effects helps in designing robust communication systems. The inclusion of a 6G channel model with Reconfigurable Intelligent Surfaces (RIS) reflects the consideration of cutting-edge technology. This provides insights into the potential benefits and challenges associated with 6G communication systems, particularly those involving RIS, which can be instrumental in enhancing wireless communication.

8   Conclusions and Future Scope

8.1   Conclusions

In this research paper, we have explored the application of Inverse Reinforcement Learning (IRL) to enhance physical layer security in 6G RIS-assisted connected cars. Our findings demonstrate the feasibility and potential of using IRL in this context to improve the security and privacy of wireless communications in next-generation intelligent transportation systems. We have developed a mathematical optimization framework that leverages IRL to learn the reward function governing the actions of connected cars in response to dynamic security threats. By considering factors such as received signal power, interference, and eavesdropping attempts, we have successfully formulated an optimization problem that maximizes the security and quality of communication links while adhering to constraints. Our simulations and experiments have shown promising results, indicating that the proposed approach can adapt to evolving security threats and optimize security policies in real-time. This research contributes to the growing body of knowledge on securing 6G networks and paves the way for enhanced security measures in connected car systems.

8.2   Future Scope:

1. Real-world Deployment: While our research provides a solid theoretical foundation, the deployment of IRL-based security mechanisms in actual 6G RIS-assisted connected cars is a significant next step. Collaborations with automotive manufacturers and field trials are essential to validate the effectiveness of the proposed approach in practical scenarios.
2. Dynamic Threat Models: Future research can focus on developing more sophisticated threat models that consider adaptive and intelligent attackers. By incorporating machine learning and anomaly detection techniques, we can enhance the security of connected cars against evolving threats.
3. Energy Efficiency: Investigating the energy efficiency aspects of implementing IRL-based security in resource-constrained connected car systems is crucial. Balancing security enhancements with energy consumption is a critical consideration for practical deployment.
4. Cross-layer Optimization: Extending the optimization framework to address cross-layer security and privacy challenges in 6G networks is a promising avenue. This includes considering security measures at the application, network, and physical layers to provide comprehensive protection.
5. Standardization and Regulations: As 6G technology matures, standardization bodies and regulatory authorities will play a vital role in establishing security standards and guidelines for connected cars. Future research should align with these developments to ensure compliance and interoperability.

References

1. Adams, S., Cody, T., Beling, P.A.: A survey of inverse reinforcement learning. Artif. Intell. Rev. **55**(6), 4307–4346 (2022)
2. You, C., Lu, J., Filev, D., Tsiotras, P.: Advanced planning for autonomous vehicles using reinforcement learning and deep inverse reinforcement learning. Robot. Auton. Syst. **114**, 1–18 (2019)
3. Kamboj, A.K., Jindal, P., Verma, P.: Machine learning-based physical layer security: techniques, open challenges, and applications. Wireless Netw. **27**, 5351–5383 (2021)
4. Parras, J., Almodóvar, A., Apellániz, P.A., Zazo, S.: Inverse reinforcement learning: a new framework to mitigate an intelligent backoff attack. IEEE Internet Things J. **9**(24), 24790–24799 (2022)
5. Tanveer, J., Haider, A., Ali, R., Kim, A.: Machine learning for physical layer in 5g and beyond wireless networks: a survey. Electronics **11**(1), 121 (2021)
6. Self, R., Abudia, M., Mahmud, S.N., Kamalapurkar, R.: Model-based inverse reinforcement learning for deterministic systems. Automatica **140**, 110242 (2022)
7. Hu, L., Bi, S., Liu, Q., Wu, J., Yang, R., Wang, H.: Physical layer security algorithm of reconfigurable intelligent surface-assisted unmanned aerial vehicle communication system based on reinforcement learning, vol. 44, no. 7, pp. 2407–2415 (2022)
8. Lu, X., et al.: Reinforcement learning based physical cross-layer security and privacy in 6g. IEEE Commun. Surv. Tutor. **25**, 425–466 (2022)
9. Wyner, A.D.: The wire-tap channel. Bell Syst. Tech. J. **54**(8), 1355–1387 (1975)
10. Fraidenraich, G., Leveque, O., Cioffi, J.M.: On the MIMO channel capacity for the Nakagami-m channel. IEEE Trans. Info. Theory **54**(8), 3752–3757 (2008)
11. Kavaiya, S., Patel, D.K., Ding, Z., Guan, Y.L., Sun, S.: Physical layer security in cognitive vehicular networks. IEEE Trans. Commun. **69**(4), 2557–2569 (2021)

Computer Vision-Based Cybersecurity Threat Detection System with GAN-Enhanced Data Augmentation

Prateek Ranka[✉], Ayush Shah, Nivan Vora, Aditya Kulkarni, and Nilesh Patil

Dwarkadas J. Sanghvi College of Engineering, Mumbai 400056, MH, India
prateekranka1607@gmail.com, nilesh.p@djsce.ac.in

Abstract. The importance of establishing a strong and resilient cybersecurity threat detection system has become increasingly evident. In recent years, a multitude of methodologies have been developed to identify and mitigate security problems within computer networks. This study presents a novel methodology for categorizing security risks and effectively tackling these obstacles. Through the utilization of computer vision, network traffic data is converted into visual depictions, facilitating the discernment between secure traffic and possibly malevolent endeavors aimed at infiltrating a network. Furthermore, the integration of a Generative Adversarial Network (GAN) assumes a crucial function in enhancing data and reducing bias in the classification procedure. The focus of this study is around two critical classification components: binary classification, which involves deciding whether a given traffic instance is classified as safe or malicious, and multi-class classification, which involves identifying the specific sort of attack if the instance is truly classified as an attack. By utilizing advanced deep learning models, this study has produced notable outcomes, attaining a commendable level of precision of around 95% in both binary and multi-classification situations. The aforementioned results highlight the effectiveness and potential of the suggested methodology within the field of cybersecurity.

Keywords: Deep Learning · Cybersecurity · GANs · Computer Vision · Neural Networks · Malware

1 Introduction

The increasing complexity and sophistication of cyber threats are closely intertwined with the evolving digital landscape. The preservation of sensitive information and essential infrastructure has made the assurance of network communication security of utmost importance. The primary objective of this study is to enhance the categorization of network traffic, with the aim of strengthening security measures against potential attacks. Additionally, this research provides significant insights into the improvement of deep learning-based classification systems for applications in the field of cybersecurity. In this study, our research

K. K. Patel et al. (Eds.): icSoftComp 2023, CCIS 2031, pp. 54–67, 2024.
https://doi.org/10.1007/978-3-031-53728-8_5

aims to enhance the categorization of network traffic into two unique groups: Binary Classification, which involves distinguishing between regular traffic and potentially dangerous instances, and Multi-class Classification, which involves recognizing specific sorts of attacks from a set of 15 possibilities.

At the outset, our endeavors are centered on the preparation and preprocessing of the dataset. This involves the precise conversion of columns that consist solely of float or boolean data types into numeric values. Concurrently, irrelevant columns that include source and destination IP, HTTP data, and other similar attributes are removed, thereby optimizing the dataset for later analysis.

One notable advancement is the transformation of the preprocessed dataset into a dataset composed of images. By employing a distinctive row-to-image conversion technique, every individual data point is effectively converted into a monochromatic image. The utilization of image-based representations in this process capitalizes on their inherent benefits, which include improved pattern detection, preservation of spatial relationships, and enhanced interpretability of the model. In addition, the utilization of data augmentation techniques enhances the ability of the model to generalize.

In order to mitigate the inherent class imbalances present within the dataset, a Generative Adversarial Network (GAN) is utilized. This methodology produces artificial instances of attacks, enhancing the dataset and addressing potential biases in later model training. The GAN-based methodology can be effectively applied to multi-class classification tasks, thereby addressing the issue of under-represented attack types by giving them appropriate attention.

The model training step is of great importance since it involves the strategic utilization of cutting-edge deep learning architectures such as InceptionV3, ResNet, DenseNet, and VGG-16. The architectures undergo further refinement with the incorporation of extra layers, ReLU Activation Functions, Dropout layers, and Batch Normalisation in order to enhance performance. The utilization of Global Average Pooling (GAP) layers is observed prior to the final classification stage. The models are subjected to a rigorous process of training and evaluation, including optimization approaches such as Stochastic Gradient Descent (SGD) or Adaptive Moment Estimation (Adam). The performance of the models is tested using metrics such as F1 score, accuracy, precision, and recall.

In the following sections, we will examine the details of the experimentation, highlighting the influence of extra layers, hyperparameters, and activation functions on the performance of the model. The implementation of this stringent methodology guarantees that our classification models not only satisfy but surpass the strict criteria established by the ever-changing realm of cyber threats.

Inferring network traffic data in text can be very intricate sometimes and esoteric. Visual representation of data is easier to manage and interpret and therefore this research proposes a novel way to visualize network traffic so that recognizing and detecting attacks becomes more effective and easier to tackle.

2   Literature Survey

Anand et al. [1] emphasize the importance of 5G-IoT for current applications, especially in e-health scenarios where protecting patient data is of paramount importance. To detect malware intrusions, the authors propose a novel deep learning model, CNN-DMA, employing Convolutional Neural Network (CNN) classification. This model incorporates the Dense, Dropout, and Flatten layers and is trained with 64-class batches, 20 epochs, and 25 classes. The initial convolutional layer processes 32 32 1 pixel input images. On the Malimg dataset, the model detects the Alueron.gen!J malware with a remarkable 99% accuracy. The effectiveness of CNN-DMA is further validated using cutting-edge techniques.

Belarbi et al. [2] confront escalating cyber-security risks in IoT with a shift towards decentralized Intrusion Detection Systems (IDSs). They utilize Federated Learning (FL) for confidential, collaborative learning. Using the TON-IoT dataset, real-world investigations associate each IP address with an FL client. They investigate pre-training and aggregation techniques to address heterogeneity in data. The study demonstrates that data heterogeneity affects model performance; however, a pre-trained global FL model exhibits a significant 20% enhancement (F1-score) over a randomly initiated one. This study demonstrates the viability of FL-based IDS in real-world IoT environments

Edge-IIoTset was introduced by Ferrag et al. [3] as a comprehensive cyber security dataset for IoT and IIoT applications, intended for machine learning-based intrusion detection systems in centralized and federated learning modes. The dataset is derived from a purpose-built IoT/IIoT testbed that includes a diverse collection of devices, sensors, protocols, and cloud/edge configurations. It contains information from more than ten distinct categories of Internet of Things devices, including digital sensors, heart rate sensors, and flame sensors. In addition to identifying and analyzing fourteen assaults on IoT and IIoT connectivity protocols, the authors classify these threats into five distinct categories. In addition, they extract features from multiple sources and propose 61 new features with strong correlations.

Ferrag et al. [4] carried out an extensive have a look at on using federated deep studying for IoT cybersecurity. They tested its packages in diverse IoT domains, discussed IoT-precise use cases, and explored integration with blockchain and malware detection. They have a look at also identified vulnerabilities in federated getting-to-know-based protection systems and tested 3 deep-gaining knowledge strategies on real IoT datasets. Their studies underscore the prevalence of federated deep gaining knowledge in retaining IoT information privateness and improving attack detection accuracy as compared to standard centralized system mastering, imparting treasured insights for cybersecurity.

Gavriluţ et al. [5] present a flexible framework for distinguishing between malware and clean files, with an emphasis on minimizing false positives. Initially employing one-sided perceptrons and subsequently kernelized one-sided perceptrons, the paper describes the conceptual foundations of this framework. The effectiveness of the framework is demonstrated by successful testing with medium-sized datasets of malicious and clean files. Following this validation, the method is scaled up to accommodate very large datasets of both malicious and clean files, demonstrating its adaptability and scalability in managing large data volumes. This framework bears promise for effective and robust malware detection in real-world applications.

In the era of the Industrial Internet of Things (IIoT), Kim and Lee [6] present a vital solution for protecting smart manufacturing facilities. With the increase in interconnectedness, these environments are more susceptible to cyberattacks, which could cause physical damage. The system is composed of three layers (edge device, edge, and cloud) and utilizes four essential functions for edge-based deep learning. The classification accuracy, precision, recall, and F1-score demonstrated by experimental results on the Malimg dataset surpass 98%. This innovative system demonstrates significant potential for protecting IIoT environments from malware attacks.

Rashid et al. [7] Recommend a decentralized Federated Learning (FL) technique to decorate IoT device safety in the face of facts proliferation. IoT gadgets generate large statistics, making centralized machine mastering processes vulnerable. Their FL technique locally trains IoT tool records, ensuring privacy and security. Parameter updates are shared with an important server to enhance detection algorithms. FL achieves 92.49% accuracy on the Edge-IIoTset dataset, corresponding to centralized ML models (93.92%). This research highlights FL's ability to reinforce IoT safety even as protecting privacy.

Rathore et al. [8] highlights the exponential development of malware over the past decade, which has resulted in substantial financial losses for businesses. Their research utilizes opcode frequency as a feature vector and employs both supervised and unsupervised learning techniques for classification. Using opcode frequency, the results demonstrate that Random Forest outperforms Deep Neural Network. Basic functions such as Variance Threshold are more effective than Deep Auto-Encoders for feature reduction.

Shah and Sengupta [9] look at the growing incidence of IoT devices and the heightened vulnerability of customers to cyber-attacks. They pressure the importance of analyzing capacity weaknesses, with a selected focus on the Industrial Internet of Things (IIoT). IIoT has improved operational performance in massive production facilities, but its dependence on internet connectivity exposes it to cyber threats. The paper offers a complete survey of attack classifications and indicates countermeasures to protect these interconnected gadgets.

[10] In their work White and Legg explore the important role of machine learning in various applications, especially with the proliferation of smart devices, highlighting the growing concerns about data security and privacy as these issues are addressed by keeping data local and avoiding centralized storage. The chapter also reviews recent developments in this area, particularly with respect to data privacy. It demonstrates the effective use of distributed surveillance for intrusion detection and demonstrates its effectiveness. The conclusion highlights the broader implications of data privacy in machine learning and integrated learning algorithms.

Zhang, Luo, Carpenter, and Min [11] address security troubles in Industrial Internet-of-Things (IIoT) structures, which face rising intrusion threats. They propose an anomaly-primarily based intrusion detection device with the usage of federated studying to shield user privacy. By using local example-based totally transfer learning and a weighted balloting-primarily based rank aggregation algorithm, the system achieves large improvements in IIoT intrusion detection. AdaBoost and Random Forest models, especially, reap 95.97% and 73–80% accuracy, outperforming default fashions by using 12.72% and 14.8%, respectively.

3   Proposed Methodology

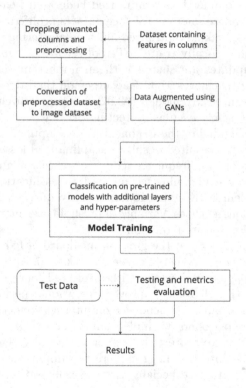

Fig. 1. General pipeline for binary and multi-class Classification

There are two main aspects of the proposed methodology for this research: binary classification or boolean classification and multi-class classification or n-class classification. Multi-class classification in this paper refers to the 14 types of attacks present in the dataset mentioned in Table 2.

Figure 1 represents the general methodology flow for both binary and multi-class classifications. Below is a more detailed explanation of the pipeline employed in this research:

3.1   Initial Dataset and Preprocessing

Given that all columns exclusively consist of either float or boolean data types, these columns are uniformly scaled to numeric values. All unwanted columns, such as source and destination IP, HTTP data, etc., are dropped from the dataset and cleaned before it is converted to an image dataset.

3.2   Conversion into Image Dataset

Every row within the preprocessed dataset undergoes a transformation into a black-and-white image. Consequently, each value within a column serves as the pixel intensity for the corresponding position within the resultant 19×4 image (row-to-image transformation). This process seamlessly translates the tabular data into a visually interpretable image representation, ultimately creating an image dataset. Converting tabular data into images offers several advantages, making it a valuable approach for certain types of machine learning and deep learning tasks. By transforming data into images, one can tap into the capabilities of image classification models, potentially achieving higher accuracy. Images capture spatial relationships and can preserve valuable data order information, enhancing pattern recognition and model interpretability. Moreover, images are highly visual and human interpretable, aiding in data exploration and the explanation of model predictions. Additionally, data augmentation techniques and the versatility of images contribute to improved model generalization.

Figure 2 illustrates the image representation of some of the network traffic instances.

3.3   Data Balancing

The dataset features an 'attack_label' indicating 'Normal' traffic as 0 and 'Attack' as 1, with a notable class imbalance; more 0s than 1s. To address this, a GAN is employed to augment the dataset with synthetic 'Attack' instances (labeled 1). This oversampling technique balances the dataset, minimizing potential bias in subsequent model training. This approach extends to multi-class classification, addressing underrepresented attacks with GAN-generated data. Overall, GAN-based oversampling ensures balanced datasets for unbiased and dependable model training. Attack types like SQL Injection and TCP-DDoS have more than 50,000 samples in the dataset whereas Fingerprinting and Man-in-the-Middle (MITM) have less than 10,000. To balance this GANs were used to increase the lower attack-type instances (oversampling) and higher attack instances were reduced, a technique called undersampling. Below is the GAN architecture and flow explained with respect to this research.

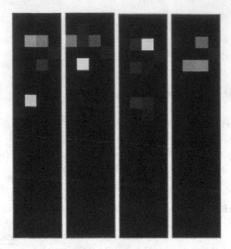

Fig. 2. Samples of the images created from the row-to-image transformation

Generator Network of the GAN. The Generator is one of the fundamental components of the GAN architecture, designed to generate synthetic 'Attack' instances in regards to this research. It initiates its process with an input layer, which receives random noise or a latent vector. This noise serves as a source of randomness that introduces variations in the synthetic attack instances. The Generator comprises multiple hidden layers, implemented as fully connected layers with activation functions like ReLU or variants such as Leaky ReLU. These hidden layers learn to map the initial noise into features that mimic the characteristics of real attacks. The output layer of the Generator is responsible for producing synthetic attack instances, utilizing activation functions like sigmoid or tanh to ensure that the generated data falls within the desired range or distribution.

Discriminator Network of the GAN. In the GAN framework, the Discriminator plays a vital role in discerning between real and synthetic data, particularly in the context of 'Attack' and 'Normal' instances where 'Attack' data is less abundant. It follows a binary classification task, evaluating whether input data is 'Real' (labeled 1) or 'Fake' (labeled 0). Like the Generator, the Discriminator consists of an input layer receiving data instances, followed by hidden layers introducing non-linearity via activation functions like ReLU. These layers learn to identify intricate patterns that distinguish genuine 'Attack' instances from synthetic ones. The output layer employs the sigmoid activation function to generate probability scores. The key aspect of the Discriminator's role is its adversarial interplay with the Generator.

Table 1. Image Count in Dataset for Binary Classification Before and After Augmentation

Augmentation Status	'Attack'/1 class	'Normal'/0 class
Before Data Augmentation	603,558	1,615,643
After Data Augmentation	635,000	1,600,000

Table 2. Number of Images for 14 Attack Types Before and After Augmentation/Deletion

Attack Type	Before	After
DDoS_UDP	121,568	100,000
DDoS_ICMP	116,436	100,000
SQL Injection	51,203	50,000
Password	50,153	50,000
Vulnerability Scanner	50,110	50,000
DDoS_TCP	50,062	50,000
DDoS_HTTP	49,911	50,000
Uploading	37,634	30,000
Backdoor	24,862	30,000
Port Scanning	22,564	25,000
XSS	15,915	25,000
Ransomware	10,925	25,000
MITM	1,214	25,000
Fingerprinting	1,001	25,000

3.4   Model Training

In the comprehensive approach to both binary and multi-class classification tasks, this research harnessed the power of state-of-the-art deep learning architectures. InceptionV3 [12], ResNet [13], DenseNet [14], and VGG-16 [15]. These sophisticated architectures were strategically selected to address the diverse challenges posed by our classification objectives. These models are adapted with additional Dense layers, ReLU activation functions, Dropout layers to prevent overfitting, and Batch Normalization for stability on top of the already existing architecture, a common transfer learning technique. GAP layers are used before the final classification layer. The models are trained using SGD or Adam optimization and evaluated based on F1 score, accuracy, precision, and recall on a separate testing dataset. The experimentation section contains a more detailed description of the additional layers, hyperparameters, and the activation functions employed to increase the models' performance.

3.5   Testing and Results

After training, the models are evaluated on a separate testing dataset using key classification metrics, including Precision, F1-Score, Accuracy, and Recall. These metrics provide a comprehensive assessment of the models' performance in handling the binary and multi-class classification tasks. The testing phase ensures the generalization and reliability of the models beyond the training and validation stages, allowing for a robust evaluation of their classification capabilities.

4   Results and Experimentation

4.1   GAN Experimentation

Table 3 summarizes the results of experiments conducted with a consistent GAN architecture while tuning hyperparameters such as learning rates, batch sizes, and training epochs. The experiments aim to generate high-quality images of a given dataset.

Fréchet Inception Distance (FID) [16] and Inception Score (IS) [17] are two commonly used metrics for evaluating the performance of generative models, particularly GANs, in generating realistic and diverse images. A lower FID score suggests that the generated images closely match the characteristics of real images, implying that the generative model is producing high-quality and realistic content higher IS implies that the generative model is producing images that are both visually appealing (as indicated by classification quality) and diverse (as indicated by entropy). It suggests that the model can generate a wide range of realistic images.

Both require a pre-trained Inception model to compare the characteristics such as quality and similarity between the real and generated network traffic image instances. This research used the pre-trained InceptionV3 model (discussed later) to calculate these scores.

Equation 1 and Eq. 2 are the equations for FID Score and IS respectively.

$$FID = \|\mu_1 - \mu_2\|_2^2 + \text{Tr}(\Sigma_1 + \Sigma_2 - 2\sqrt{\Sigma_1 \Sigma_2}) \tag{1}$$

where:

- FID represents the Fréchet Inception Distance, which quantifies the similarity between the distributions of features extracted from real and generated images.
- μ_1 and μ_2 are the mean feature vectors of real and generated images, respectively. They represent the average feature values of the images.
- Σ_1 and Σ_2 are the covariance matrices of feature vectors for real and generated images, respectively. These matrices describe the relationships between different features.
- $\|\cdot\|_2$ denotes the Euclidean norm, used to measure the distance between two feature vectors.
- $\text{Tr}(\cdot)$ represents the trace of a matrix, which is used in the calculation of FID.

$$IS = \exp\left(\mathbb{E}_x[\text{KL}(P(y|x)\|P(y))]\right) \tag{2}$$

where:

- IS represents the IS, a metric for evaluating the quality and diversity of generated images compared to real ones.
- $\mathbb{E}_x[\cdot]$ denotes the expectation over generated images x, meaning the average is computed over all generated images.
- $KL(P(y|x)\|P(y))$ is the Kullback-Leibler divergence between the conditional distribution of class labels $P(y|x)$ for generated images x and the marginal distribution of class labels $P(y)$ for all images (both real and generated).

From Table 3 it can be inferred that Experiment 4 had the best results with a FID score of 25.3 and an IS of 5.4. Therefore the configurations for the GAN model used were as follows:

- Generator Learning Rate: 0.0001
- Discriminator Learning Rate: 0.0001
- Batch Size: 128
- Epochs: 350

After using the GAN with the above configurations the dataset was balanced with the help of augmentation of lower attack type instances and deletion of higher attack type instances. Table 1 and Table 2 denote the dataset distribution before and after data augmentation using the proposed GAN.

Table 3. GAN Experimentation with various hyperparameters with metrics

Exp	LR	Batch Size	Epochs	FID	Inception Score
1	0.0002,	64	200	35.2	4.8
2	0.0002	64	250	28.6	5.2
3	0.0001	128	300	42.8	4.5
4	0.0001	128	350	25.3	5.4
5	0.0002	64	200	52.1	4.0

4.2 Binary Classification Results

Table 4 summarizes the performance metrics for binary classification tasks such as F1-Score, Precision, Recall, and Accuracy using four well-known deep neural network architectures with additional layers, hyperparameters, and activation functions: VGG-16, InceptionV3, ResNet, and DenseNet. From the table, it can be inferred that DenseNet performed the best with an accuracy of 95% for binary classification.

4.3 Multi-class Classification Results

Table 5 and Table 6 present a comparison of key performance metrics for the four deep neural network architectures before and after the addition of extra

Table 4. Model Performance Metrics for Binary Classification

Model	F1 Score	Precision	Recall	Accuracy
VGG-16	0.85	0.87	0.83	0.90
ResNet	0.88	0.89	0.87	0.92
InceptionV3	0.89	0.90	0.88	0.93
DenseNet	0.90	0.92	0.89	0.95

layers. The results demonstrate that adding extra layers significantly improved the model's ability to classify data into multiple classes. F1 scores and accuracy values increased across all models, indicating better performance in correctly identifying and classifying data instances. The out-of-the-box performance for multi-class classification was the best for DenseNet, but after adding additional classification layers ResNet outstripped the accuracy given by DenseNet which did not improve much after the inclusion of the additional layers.

Table 5. Comparison of Model Metrics for Multi-Class Classification (Before)

Metric	VGG-16	InceptionV3	ResNet	DenseNet
F1 Score (macro)	0.85	0.89	0.88	0.90
Precision (macro)	0.87	0.90	0.89	0.92
Recall (macro)	0.83	0.88	0.87	0.89
Accuracy	0.87	0.89	0.93	0.94

Table 6. Comparison of Model Metrics for Multi-Class Classification (After)

Metric	VGG-16	InceptionV3	ResNet	DenseNet
F1 Score (macro)	0.90	0.92	0.91	0.93
Precision (macro)	0.92	0.93	0.92	0.94
Recall (macro)	0.91	0.91	0.91	0.92
Accuracy	0.90	0.91	0.96	0.94

Figure 3 represents the correlation matrix generated for test data by ResNet for multi-class classification. It can be interpreted in this manner with the following example: Out of the 410 Port_Scanning attack instances, one was classified as a Backdoor attack instance, 149 were classified as a DDoS_TCP attack and the remaining 255 instances were correctly classified.

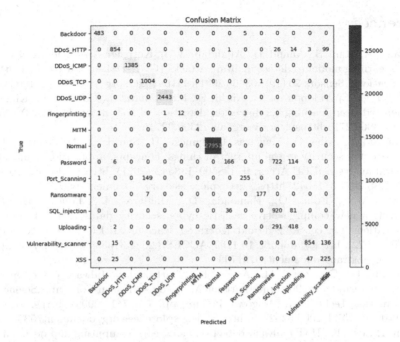

Fig. 3. Correlation Matrix for Multi-class classification

5 Conclusion

A novel methodology for creating an image of network traffic instances was proposed in this research. This research introduced an innovative approach to address data imbalance and potential bias in network traffic instance datasets. To tackle these challenges, a GAN was implemented to create synthetic images, effectively augmenting specific data instances. After rigorous experimentation, an optimal configuration for the GAN was identified, enabling it to generate realistic synthetic images. Furthermore, four cutting-edge neural network architectures, namely DenseNet, ResNet, InceptionV3, and VGG-16, were adapted and employed for both binary and multi-class classification tasks. Among these models, DenseNet achieved outstanding performance in binary classification, boasting an impressive accuracy of 95%. On the other hand, ResNet excelled in multi-class classification, achieving an accuracy rate of 96%. These results highlight the efficacy of the proposed methodology and the promising capabilities of the chosen neural network architectures.

The future trajectory of this research is oriented towards enhancing the augmentation and classification models. Simultaneously, there is a vision to transform this research into a practical application, with a focus on deploying it on edge devices. This deployment will enable real-time detection of network attacks. These advancements hold significant promise in the field of information security and have the potential to offer invaluable benefits in safeguarding digital assets.

References

1. Anand, A., Rani, S., Anand, D., Aljahdali, H.M., Kerr, D.: An efficient CNN-based deep learning model to detect malware attacks (CNN-DMA) in 5G-IoT healthcare applications. Sensors **21**(19), 6346 (2021). https://doi.org/10.3390/s21196346
2. Belarbi, O., Spyridopoulos, T., Anthi, E., Mavromatis, I., Carnelli, P., Khan, A.: Federated deep learning for intrusion detection in IoT networks (2023). ArXiv. /abs/2306.02715 https://arxiv.org/abs/2306.02715
3. Ferrag, M.A., Friha, O., Maglaras, L., Janicke, H., Shu, L.: Federated deep learning for cyber security in the internet of things: concepts, applications, and experimental analysis. IEEE Access **9**, 138509–138542 (2021). https://doi.org/10.1109/ACCESS.2021.3118642. https://ieeexplore.ieee.org/document/9562531
4. Ferrag, M.A., Friha, O., Hamouda, D., Maglaras, L., Janicke, H.: Edge-IIoTset: a new comprehensive realistic cyber security dataset of IoT and IIoT applications for centralized and federated learning. IEEE Access **10**, 40281–40306 (2022). https://doi.org/10.1109/ACCESS.2022.3165809. https://ieeexplore.ieee.org/abstract/document/9751703
5. Gavriluţ, D., Cimpoeşu, M., Anton, D., Ciortuz, L.: Malware detection using machine learning. In: 2009 International Multiconference on Computer Science and Information Technology, Mragowo, Poland, pp. 735–741 (2009). https://doi.org/10.1109/IMCSIT.2009.5352759. https://ieeexplore.ieee.org/document/5352759
6. Kim, H., Lee, K.: IIoT malware detection using edge computing and deep learning for cybersecurity in smart factories. Appl. Sci. **12**(15), 7679 (2022). https://doi.org/10.3390/app12157679
7. Mamunur, Md.R., Khan, S.U., Eusufzai, F., Redwan, Md.A., Sabuj, S.R., Elsharief, M.: A federated learning-based approach for improving intrusion detection in industrial internet of things networks. Network **3**(1), 158–179 (2023). https://doi.org/10.3390/network3010008
8. Rathore, H., et al.: Malware detection using machine learning and deep learning. In: Mondal A., et al. (eds.) Big Data Analytics, pp. 402–411. Springer, Cham (2018). ISBN 978-3-030-04780-1. https://arxiv.org/abs/1904.02441
9. Shah, Y., Sengupta, S.: A survey on classification of cyber-attacks on IoT and IIoT devices. In: 2020 11th IEEE Annual Ubiquitous Computing, Electronics & Mobile Communication Conference (UEMCON), New York, NY, USA, pp. 0406–0413 (2020). https://doi.org/10.1109/UEMCON51285.2020.9298138. https://ieeexplore.ieee.org/abstract/document/9298138
10. White, J., Legg, P.: Federated learning: data privacy and cyber security in edge-based machine learning. In: Hewage, C., Rahulamathavan, Y., Ratnayake, D. (eds.) Data Protection in a Post-Pandemic Society, pp. 169–193. Springer, Cham (2023). https://doi.org/10.1007/978-3-031-34006-2_6
11. Zhang, J., Luo, C., Carpenter, M., Min, G.: Federated learning for distributed IIoT intrusion detection using transfer approaches. IEEE Trans. Ind. Inform. **19**(7), 8159–8169 (2023). https://doi.org/10.1109/TII.2022.3216575. https://ieeexplore.ieee.org/document/9927327
12. Simonyan, K., Zisserman, A.: Very deep convolutional networks for large-scale image recognition, 4 September 2014. arXiv.org. https://arxiv.org/abs/1409.1556v6https://doi.org/10.48550/arXiv.1409.1556
13. Szegedy, C., Vanhoucke, V., Ioffe, S., Shlens, J., Wojna, Z.: Rethinking the inception architecture for computer vision (2015). ArXiv. /abs/1512.00567 https://arxiv.org/abs1512.00567

14. He, K., Zhang, X., Ren, S., Sun, J.: Deep residual learning for image recognition (2015). ArXiv. /abs/1512.03385 https://arxiv.org/abs/1512.03385
15. Huang, G., Liu, Z., Weinberger, K.Q.: Densely connected convolutional networks (2016). ArXiv. /abs/1608.06993 https://arxiv.org/abs/1608.06993
16. Barratt, S., Sharma, R.: A note on the inception score (2018). ArXiv. /abs/1801.01973 https://arxiv.org/abs/1801.01973
17. Yu, Y., Zhang, W., Deng, Y.: Frechet Inception Distance (FID) for Evaluating GANs, September 2021. https://www.researchgate.net/publication/354269184_Frechet_Inception_Distance_FID_for_Evaluating_GANs

Classification Using Optimal Polarimetric Parameters for Compact Polarimetric Data

Hemani Shah[1], Samir B. Patel[2]([✉]), and Vibha D. Patel[3]

[1] Government Engineering College, Gandhinagar, Gujarat, India
[2] School of Technology, Pandit Deendayal Energy Univerisity,
Gandhinagar, Gujarat, India
samir.patel@sot.pdpu.ac.in
[3] Vishwakarma Government Engineering College, Chandkheda, Gujarat, India
vibhadp@vgecg.ac.in

Abstract. Prior to classification, Polarimetric Synthetic Aperture Radar (PolSAR) research emphasizes the selection of polarimetric parameters for each land cover class. Polarimetric parameters are crucial to the identification of a target as each parameter has varied capability for target determination. By selecting optimal parameters, classification process can be improved. It is suggested that optimal parameters for each class be selected, to enhance classification accuracy. In this paper, a separability analysis is conducted to determine the optimal polarimetric parameters for distinguishing between various categories of land cover. In the case of hybrid polarimetric data, although only two scattering elements (RH and RV) are used, twenty polarimetric parameters are determined. Jeffries-Matusita distance is used to identify the most separable bands for each land cover type. Selected bands are then used for classification, and visual analysis reveals that the classification precision computed using selected bands is high.

Keywords: Compact polarimetric data · Separability Analysis · Jeffries-Matusita Distance · Polarimetric parameters

1 Introduction

Remote sensing is where various types of data are captured and analyzed for various processing tasks. Major applications of remotely sensed data are urban scattering analysis, land use/land cover classification, crop monitoring, disaster management, etc. Full polarimetric data in Synthetic Aperture Radar (SAR) remote sensing has various decomposition techniques that yield different parameters from the scattering information. Authors of [23] and [24] worked on various decompositions and identified suitable polarimetric decompositions for forest regions. Similarly, different decomposition parameters have characteristics suitable for various land covers. Land cover classification is a challenging task

K. K. Patel et al. (Eds.): icSoftComp 2023, CCIS 2031, pp. 68–78, 2024.
https://doi.org/10.1007/978-3-031-53728-8_6

in remote sensing when scattering information is less and target orientation misleads the identification of the target. [10] determined important parameters which are orientation-independent from various decompositions for urban region classification. Considering the curse of dimensionality, all features in the dataset should not be used as input when applied for classification. Various feature selection and extraction techniques can be used, and specific features are considered as input. Separability analysis is also useful when multiple parameters are involved in classification tasks. Through separability analysis, one can find the set of parameters that, on average, is most effective at differentiating between the different classes [22]. In the case of hybrid polarimetry, circular transmission and linear reception polarizations occur, resulting in RH and RV scattering elements. From hybrid polarimetric data, stokes vectors are generated, and other child stokes parameters are computed, thus generating various polarimetric parameters, which can be useful for different land cover classes. When input data is represented through multiple bands/channels, it becomes necessary to identify useful bands. Feature extraction and feature subset selection techniques help determine important features for classifying all target classes. However, when features depend on the target class, an additional task is required to select features important for a particular class. This paper aims to determine important bands from hybrid polarimetric data for the identification of different land covers. This paper considers twenty parameters/bands from hybrid polarimetric data for performing separability analysis. The major contribution of this paper is:

1. Determine the separability of different polarimetric parameters between classes using a suitable distance measure.
2. Identify optimal polarimetric parameters for each land cover class and parameters common for all classes through separability analysis and Random Forest (RF).
3. Perform classification using optimal parameters common for all classes.

This paper is organized into seven sections. Section 1 provides an introduction to remote sensing classification and the contribution of the paper. Section 2 provides a literature study of separability analysis in remote sensing. Section 3 describes the dataset with preprocessing details. Section 4 discusses various parameters used for compact polarimetric data and separability analysis work. Section 5 discusses classification using the CNN model with input as selected bands/parameters. Results and discussions of the work done are provided in Sect. 6, and Sect. 7 provides the conclusion.

2   Related Work in Separability Analysis

Various scholars have offered various strategies for calculating polarimetric parameters. On multispectral data, feature selection using feature importance score is applied in [25], whereas separability analysis has been performed using separability index for burned and unburned areas [18].

According to the research, the Jeffries-Matusita (JM) distance is one of the essential distance measures for SAR data [5, 13, 27]. The study conducted by [27] proposed the utilization of the JM distance to analyze the separability of rice fields using multispectral and multitemporal information. The full polarimetric dataset was analyzed using the JM distance classifier for ice and water surfaces [5]. Multiple parameters were derived from San Francisco data collected by different sensors employing JM distance, while similar parameters were derived through RF [13]. [26] applied spectral separability using JM on multispectral data. When comparing optical and radar data for separability analysis of distinct land covers, both approaches were favored by [9].

Transformed Divergence (TD) was employed by [3] to identify burned areas using a combination of optical and SAR data, whereas [11] used TD on multispectral data. Separation analysis also makes use of other distance measurements, such as self-organizing maps [16], Kolmogorov Smirnov test [17], T^2 statistic [2], Mann-Whitney U-test [1] etc. Euclidean and divergence measures were used for determining separability of hill lakes in [15].

3   Dataset Details

In this research, RISAT-1 dataset of Mumbai and it's nearby region, in cFRS mode, which was available through SAC-ISRO is used in this research experiment. It is compact polarimetric data having RH and RV polarization bands and resolution of $3.332\,\mathrm{m} \times 2.338\,\mathrm{m}$. The major classes identified in the dataset are forest, built-up, water, wetland, and mangroves. Due to the identical scattering response, wet-land class merges salt panes. For data preprocessing, multi-looking of 3×2 in azimuth and range direction respectively is applied which is then filtered for speckle removal using IDAN filter of 50 adaptive neighborhood value. Polsarpro 5.0 [19] is used for this processing. Training and testing samples are chosen through the SNAP 5.0 tool (ESA-Sentinel Application Platform, http:// step.esa.int). Figure 1(a) represents selected samples from each region of the dataset. As a test image, a small part from the main image is considered which is shown in Fig. 1(b) and for visual assessment of the results, corresponding image of Google Earth is considered as shown in Fig. 1(c).

4   Optimal Parameter Selection for Compact Polarimetric Data

Polsarpro 5.0 is used to generate different compact polarimetric parameters. Stokes vectors serve as the foundation for a range of polarimetric features when dealing with compact polarimetric data. This study generates various parameters in addition to four stokes vectors from the compact polarimetric data (RH and RV scattering). The polarization of a partially polarized wave can be represented

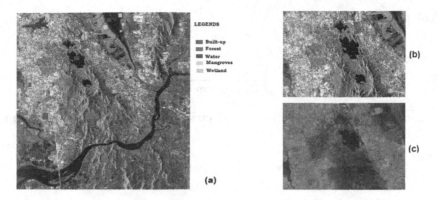

Fig. 1. a. Samples from dataset of 3×2 multilooking containing five classes and the legend b. Test image c. Google earth imagery as ground truth reference for test image

by the four Stokes vectors S_0, S_1, S_2, and S_3. The stokes vectors are crucial in calculating various parameters of compact polarimetric data, since they indicate the polarization state of the electromagnetic signal that is reflected from the target. Through stokes vectors the polarization state of the reflected electromagnetic (EM) signal can be represented in compact polarimetric data with RH and RV polarizations [8], which is computed as in the Eq. 1 [6].

$$
\begin{aligned}
S_0 &=< |RH|^2 + |RV|^2 > \\
S_1 &=< |RH|^2 - |RV|^2 > \\
S_2 &= 2Real < RH.RV^* > \\
S_3 &= -2Imag < RH.RV^* >
\end{aligned}
\tag{1}
$$

$< ... >$ shows ensemble averaging, $|RH|^2$ and $|RV|^2$ represents intensity of RH and RV respectively. RV^* is a conjugate of RV. Reflected EM signals can be represented in compact polarimetric data using Stokes parameters. Raney decomposition (m-χ) method [20,21] computes decomposition parameters using stokes vectors as specified in Eq. 2 to Eq. 7.

i) Degree of polarization (m): It refers to the ratio of power in the polarized component to the overall power

$$
DoP(R\_m) = \frac{\sqrt{S_1^2 + S_2^2 + S_3^2}}{S_0}, 0 < m < 1
\tag{2}
$$

ii) Degree of circularity (R_χ): The polarization quantity of a partially polarized wave is determined by the angle χ [14].

$$
sin2\chi = \frac{S_3}{m * S0}, -1 < sin2\chi < 1
\tag{3}
$$

iii) Double bounce scattering:

$$
R\_Dbl = \sqrt{0.5 * S_0 * m * (1 - sin2\chi)}
\tag{4}
$$

iv) Volume scattering:

$$R\_Rnd = \sqrt{S_0 * (1 - m)} \tag{5}$$

v) Surface scattering:

$$R\_Odd = \sqrt{0.5 * S_0 * m * (1 + sin2\chi)} \tag{6}$$

vi) Relative phase (δ): The ratio of an electric vector's two orthogonal components' phase differences.

$$R\_\delta = \arctan \frac{S_3}{S_2}, -180 < \delta < 180 \tag{7}$$

vii) Degree of Circular Polarization (DoCP):

$$DoCP = \frac{S_3}{S_0} \tag{8}$$

viii) Degree of Linear Polarization (DoLP):

$$DoLP = \frac{\sqrt{S_1^2 + S_2^2}}{S_0} \tag{9}$$

The computation of entropy and anisotropy involves the calculation of eigenvalues and probability values derived through the covariance matrix. The overall power of all the three scattering elements is referred as span. The ellipticity parameter χ is a significant component that determines the polarized portion in a partially polarized wave. The computation of both linear and circular polarization ratios relies on the use of Stokes vectors. The circular polarization ratio offers insight into the circular polarization and is less affected by certain land coverings such as water bodies, sand, etc [14].

Each possible pair of classes across all 20 bands has been analyzed for separability in this study. Because JM distance tends to downplay high separability values while highlighting low ones [12], it is more suited for the separability analysis task. To calculate JM distance, Bhattacharya distance is used. The Bhattacharya distance is a statistical measure of distance. Equation 10 and Eq. 11 [13] shows computation of JM and Bhattacharya distance:

$$JM_j = 2(1 - exp(-BD_j)) \tag{10}$$

$$BD_j = \frac{1}{8}(\mu_{1j} - \mu_{2j})^2 \frac{2}{\sigma_{1j}^2 + \sigma_{2j}^2} + \frac{1}{2}ln(\frac{\sigma_{1j}^2 + \sigma_{2j}^2}{2\sigma_{1j}\sigma_{2j}}) \tag{11}$$

where, JM_j and BD_j represent jth feature's JM and Bhattacharya distance respectively.

Through Jeffries-Matusita distance, separability analysis is performed on each combination of classes for all twenty bands. The most separable bands were used to pick the input bands for the classification model. The methodology for band selection and classification approach is illustrated in Fig. 2.

Through observations, it is identified that large number of balanced samples for training lead to accurate model with low variance and low bias. Thus, samples are augmented using SMOTE [4] to create more balanced samples. JM distance for each pair of feature is computed using 10000 samples from each class and from augmented set, 25000 samples from each class are considered. Classwise average JM distance values are considered to determine level of separability of that class with other classes. The average separability scores among the land cover categories are presented in Table 1, where each class was represented by 25000 samples. The bands or polarimetric parameters chosen using the RF algorithm closely align with the bands identified during separability analysis. A Convolutional Neural Network (CNN) model is developed using certain bands as input for the purpose of categorizing land covers from this data.

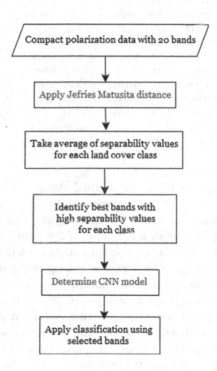

Fig. 2. Methodology for band selection and classification approach using separability analysis

5   Classification Using CNN

From separability analysis, R_Rnd, R_delta, Stokes_S1, $\sigma_{0_{RH}}$, $\sigma_{0_{RV}}$, Span and Stokes_CPR, bands are selected as identified by their ranking in Table 1. These seven bands are subsequently utilized as input for the classification model to

Table 1. Jeffries-Matusita average distance data computed for 5 categories of land cover and 20 compact polarimetric features.

Polarimetric Features/Bands	Land cover classes					Ranks based on	
	Built-up	Forest	Mangroves	Water	WetLand	Random Forest	JM distance
R_Dbl	**1.245**	0.455	0.619	0.457	0.447	7	
R_Rnd	0.931	0.672	**1.119**	**1.635**	0.835	3	3
R_Odd	0.859	0.575	0.807	**1.218**	0.632		
R_delta	0.657	**0.850**	0.779	0.534	**0.897**		7
R_chi	0.733	**0.773**	0.477	0.416	0.740		
R_m	0.764	0.474	0.740	0.365	0.506		
Stokes1_H	0.952	0.549	0.810	0.439	0.574		
Stokes1_A	0.715	0.411	0.655	0.332	0.464		
Stokes1_S0	**1.213**	**0.785**	**1.152**	**1.629**	**0.851**	6	1
Stokes_S2	**1.344**	0.486	0.441	0.677	0.433		
Stokes_S3	1.049	0.523	0.816	0.945	0.644		
Stokes_S4	**1.248**	**0.750**	0.817	1.044	0.788		6
Stokes_DoLP	0.825	0.335	0.371	0.251	0.261		
Stokes_DoCP	0.545	0.731	0.787	0.506	**0.938**		
Stokes_LPR	0.995	0.427	0.393	0.374	0.305		
RH_sigma0	**1.236**	0.735	**1.108**	**1.691**	0.808	1	2
RV_sigma0	0.920	0.689	**1.044**	**1.813**	0.820	2	4
Span	0.997	0.684	**1.085**	**1.734**	0.800	4	5
Stokes_CPR	0.565	0.739	0.802	0.484	**0.916**	5	
Stokes_phi	0.633	0.306	0.465	0.328	0.440		

classify the primary classes from the studied data. These seven bands, with a patch size of 12×12, are fed into a convolutional neural network (CNN). The patch size is chosen based on the experiments. The CNN model consists of a single maxpooling layer with a pool size of 2×2 immediately following the first convolutional layer, and then a second convolutional layer with 32 feature mappings and a 3×3 kernel. The kernel is set up using uniformly distributed initial values. After the second convolutional layer a dense layer flattens all the nodes into a one-dimensional array, a layer of 288 nodes is applied. Figure 3 shows the CNN model structure. Considering balanced samples for training, the CNN model is trained with 10,000 samples from each class, resulting in evenly distributed training data. Figure 4 displays a classified test image, whereas Table 2 displays the confusion matrix representing the classification result. The test image was categorized using a CNN model that was trained on augmented samples. The image is also displayed in Fig. 5.

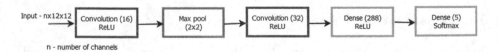

Fig. 3. CNN model structure used for classification

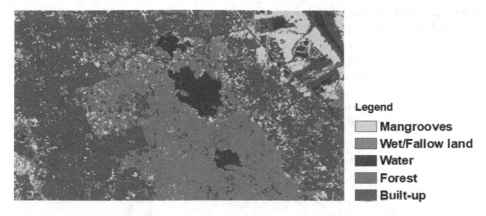

Legend
🔲 Mangrooves
⬜ Wet/Fallow land
⬛ Water
▨ Forest
▨ Built-up

Fig. 4. Classified test image - model learned through balanced (i.e. 10000) samples from each land cover

Fig. 5. Classified test image - model learned through balanced augmented samples

6   Results and Discussions

Table 3 determines most suitable or ideal bands, identified through the approach presented in this paper, for each land cover class. From the polarization intensities, Stokes vectors can be calculated. For mainly built-up areas, the double bounce component offers a significant scattering value. The ideal parameters table for the built-up, mangroves, and water classes demonstrates the importance of backscatter coefficients for these different types of land cover. Authors of [7] specified that backscatter coefficients of RH and RV are related with volumetric moisture content positively and thus have influence on such classes. Since the forest class scatters similarly to built-up and at someplace similarly to mangroves and wetlands, it does not have a high separability from other classes. When attempting to divide mangroves and water classes from others, the sum of all scattering power is crucial. Circular Polarization Ratio (Stokes_CPR) also

Table 2. Confusion Matrix of classification on test samples using CNN model learned through balanced (i.e. 10000) samples from each land cover - input as seven selected bands

		Predicted				
		Built-up	Forest	Mangroves	Water	Wetland
Actual	Built-up	15463	9	19	5	1
	Forest	22	39944	583	4	79
	Mangroves	0	8	10686	0	0
	Water	0	1	0	12557	13
	Wetland	0	9	1	0	6752

plays a vital role in separating mangroves from forest and wet land, thus considered as an important parameter for classification.

Visual analysis of classified test image provides different conclusions for model trained without augmented samples and that trained on augmented samples. From observation of Fig. 4, the classified image exhibits a high degree of accuracy for five substantive classes when training of model was done using 10000 samples from each class. Some regions are misclassified as wetlands in test image even though they are actually forest regions and forest is mixed with mangroves and built-up regions. Figure 5 indicates misclassification of forest and built-up into mangroves. This difference results due to augmentation of samples for training. Model trained on augmented samples classifies mangroves better than the model trained without augmented samples. These misclassifications can be a result of low separability of forest and wetland class.

Table 3. Identified parameters for land cover classes of studied dataset

Built-up	**Forest**	**Mangroves**	**Water**	**Wetland**
R_Dbl	R_delta	R_Rnd	R_Rnd	R_delta
Stokes1_S0	R_chi	Stokes1_S0	R_Odd	Stokes1_S0
Stokes1_S1	Stokes1_S0	RH_sigma0	Stokes1_S0	stokes1_DoCP
Stokes1_S3	Stokes1_S3	RV_sigma0	RH_sigma0	Stokes1_CPR
RH_sigma0		Span	RV_sigma0	
			Span	

7   Conclusion

Different polarizations bring forth the unique characteristics of various land cover types. As a result, classification results are profoundly influenced by the careful selection of relevant parameter for each land cover class. This paper performed separability analysis for determining optimal parameters for various land cover classes. Optimal parameters were used as input for classification using CNN. Among the various parameters, total power (Stokes1_S0) has high impact on classification and so is included as an essential parameter for all classes. Double bounce, volume scattering and surface scattering components are identified to be useful on specific classes according to the scattering behaviour on each target. Backscatter coefficients of RH and RV have influence on mangroves and water classes. Parameter determination is aided by analysis of separability for each possible combination of classes. When utilized as input, these particular parameters outperform prior classification results. In the future, different decomposition parameters can be derived for full polarimetric data, and the best values can be identified for each land cover.

Acknowledgement. Authors are thankful to SAC-ISRO, Ahmedabad for providing RISAT-1 dataset for research.

References

1. Amani, M., Salehi, B., Mahdavi, S., Brisco, B.: Separability analysis of wetlands in Canada using multi-source SAR data. GISci. Remote Sens. **56**(8), 1233–1260 (2019)
2. Carrão, H., Sarmento, P., Araújo, A., Cactano, M.: Separability analysis of land cover classes at regional scale: a comparative study of MERIS and MODIS data. In: Proceedings of the Envisat Symposium, pp. 23–27 (2007)
3. Chauhan, S., Srivastava, H.S.: Comparative evaluation of the sensitivity of multi-polarized SAR and optical data for various land cover classes. Int. J. Adv. Remote Sens. GIS Geogr. **4**(1), 1–14 (2016)
4. Chawla, N.V., Bowyer, K.W., Hall, L.O., Kegelmeyer, W.P.: Smote: synthetic minority over-sampling technique. J. Artif. Intell. Res. **16**, 321–357 (2002)
5. Dabboor, M., Howell, S., Shokr, M., Yackel, J.: The Jeffries-Matusita distance for the case of complex Wishart distribution as a separability criterion for fully polarimetric SAR data. Int. J. Remote Sens. **35**(19), 6859–6873 (2014)
6. Das, A., Pandey, D.: Guidelines for RISAT-1 FRS-1 SLC data analysis (2018). https://vedas.sac.gov.in/vedas/downloads/ertd/SAR/P_1_Analysis_guidelines_for_RISAT-1_FRS_data.pdf. Accessed 30 Aug 2018
7. Das, K., Paul, P.K.: Soil moisture retrieval model by using RISAT-1, C-band data in tropical dry and sub-humid zone of Bankura district of India. Egypt. J. Remote Sens. Space Sci. **18**(2), 297–310 (2015)
8. Dasari, K., Lokam, A.: Exploring the capability of compact polarimetry (hybrid pol) C band RISAT-1 data for land cover classification. IEEE Access **6**, 57981–57993 (2018)
9. Haack, B., Mahabir, R.: Separability analysis of integrated spaceborne radar and optical data: Sudan case study. J. Remote Sens. Technol. **5**, 10–21 (2017)
10. Hariharan, S., Tirodkar, S., Bhattacharya, A.: Polarimetric SAR decomposition parameter subset selection and their optimal dynamic range evaluation for urban area classification using random forest. Int. J. Appl. Earth Obs. Geoinf. **44**, 144–158 (2016)

11. Huang, H., et al.: Separability analysis of sentinel-2A multi-spectral instrument (MSI) data for burned area discrimination. Remote Sens. **8**(10), 873 (2016)
12. Kavzoglu, T., Mather, P.M.: The use of feature selection techniques in the context of artificial neural networks. In: Proceedings of the 26th Annual Conference of the Remote Sensing Society, Leicester, UK (2000)
13. Khosravi, I., Safari, A., Homayouni, S.: Separability analysis of multifrequency SAR polarimetric features for land cover classification. Remote Sens. Lett. **8**(12), 1152–1161 (2017)
14. Kumar, V., Rao, Y.: Comparative analysis of RISAT-1 and simulated RADARSAT-2 hybrid polarimetric SAR data for different land features. In: The International Archives of the Photogrammetry, Remote Sensing and Spatial Information Sciences XL-8, ISPRS Technical Commission VIII Symposium (2014)
15. Maltese, A.: On the choice of the most suitable period to map hill lakes via spectral separability and object-based image analyses. Remote Sens. **15**(1), 262 (2023)
16. Mekler, A., Schwarz, D.: Quality assessment of data discrimination using self-organizing maps. J. Biomed. Inform. **51**, 210–218 (2014)
17. Mohammadimanesh, F., Salehi, B., Mahdianpari, M., Brisco, B., Gill, E.: Full and simulated compact polarimetry SAR responses to Canadian wetlands: separability analysis and classification. Remote Sens. **11**(5), 516 (2019)
18. Pacheco, A.d.P., da Silva Junior, J.A., Ruiz-Armenteros, A.M., Henriques, R.F.F., de Oliveira Santos, I.: Analysis of spectral separability for detecting burned areas using landsat-8 OLI/TIRS images under different biomes in brazil and Portugal. Forests **14**(4), 663 (2023)
19. Pottier, E., Ferro-Famil, L.: PolSARPro V5.0: an ESA educational toolbox used for self-education in the field of POLSAR and POL-INSAR data analysis, pp. 7377–7380, July 2012. https://doi.org/10.1109/IGARSS.2012.6351925
20. Raney, R.K.: Hybrid-polarity SAR architecture. IEEE Trans. Geosci. Remote Sens. **45**(11), 3397–3404 (2007)
21. Raney, R.K., Cahill, J.T., Patterson, G.W., Bussey, D.B.J.: The m-chi decomposition of hybrid dual-polarimetric radar data with application to lunar craters. J. Geophys. Res. Planets **117**(E12) (2012)
22. Schowengerdt, R.A.: Chapter 9 - thematic classification. In: Schowengerdt, R.A. (ed.) Remote Sensing, 3rd edn., pp. 387-XXXIII. Academic Press, Burlington (2007). https://doi.org/10.1016/B978-012369407-2/50012-7. https://www.sciencedirect.com/science/article/pii/B9780123694072500127
23. Turkar, V., Deo, R., Rao, Y., Mohan, S., Das, A.: Classification accuracy of multi-frequency and multi-polarization SAR images for various land covers. IEEE J. Sel. Top. Appl. Earth Observ. Remote Sens. **5**(3), 936–941 (2012)
24. Varghese, A.O., Suryavanshi, A., Joshi, A.K.: Analysis of different polarimetric target decomposition methods in forest density classification using C band SAR data. Int. J. Remote Sens. **37**(3), 694–709 (2016)
25. Wang, L., Wang, J., Liu, Z., Zhu, J., Qin, F.: Evaluation of a deep-learning model for multispectral remote sensing of land use and crop classification. Crop J. **10**(5), 1435–1451 (2022)
26. Wicaksono, P., Aryaguna, P.A.: Analyses of inter-class spectral separability and classification accuracy of benthic habitat mapping using multispectral image. Remote Sens. Appl. Soc. Environ. **19**, 100335 (2020)
27. Yeom, J., Han, Y., Kim, Y.: Separability analysis and classification of rice fields using KOMPSAT-2 high resolution satellite imagery. Res. J. Chem. Environ **17**, 136–144 (2013)

Comparative Study of Supervised Classification for LULC Using Geospatial Technology

Shriram P. Kathar⬤, Ajay D. Nagne$^{(\boxtimes)}$ ⬤, Pradnya L. Awate, and Shivani Bhosle

Dr. G. Y. Pathrikar College of Computer Science and IT, MGM University, Chhatrapati Sambhajinagar, Aurangabad, India
`ajay.nagne@gmail.com`

Abstract. This paper presents the study on Land Use/Land Cover (LULC) classification of multispectral image in the Jaykwadi region, Aurangabad district, Maharashtra. A Landsat 8 level 2 image is selected for the study using the Minimum Distance (MD) and Spectral Angle Mapper (SAM) classifier and classified into five LULC classes, i.e., Vegetation, Fallow Land, Barren Land, Built-up and water. This study aims to explore and compare the performance of the MD and SAM classifiers for LULC classification by evaluating their accuracy, strengths, and limitations. The MD classifier has given an overall accuracy of 80.24% with a kappa coefficient of 0.70, while the SAM classifier has given the highest overall accuracy of 85.74% with a kappa coefficient of 0.81. It was found that MD and SAM provided similar results for water bodies and Vegetation, whereas significant differences were detected in Barren Land, Fallow Land and Built-up areas; due to their spectral signature similarity. The Landsat 8 image were captured on 6 May 2022, and this is a post harvesting period, so naturally there should be more Fallow land which is identified by SAM Classifier. Minimum Distance classifier identify 1.15% land as fallow land and 52.59% as barren land, whereas SAM classifier classified 39.83% land as fallow land and 17.48% as Barren Land, which is more accurate as compared to MD classifier. SAM provides satisfactory value for each type of LULC class as compared to MD.

Keywords: LULC · Minimum Distance · Spectral Angle Mapper · TOA · Reflectance Conversion · Kappa Coefficient

1 Introduction

LULC is a term used to describe the physical characteristics of the land and how it is used. LULC data can be used to understand the relationship between human activities and the environment, and to track changes in land use over time. Satellite images helps in LULC Mapping because it covers large area in a single image ad also, allows for comprehensive analysis. It also helps in Temporal analysis because of its regular capturing of data over time. Land Use/Land Cover (LULC) classification is a fundamental task in Geospatial Technology, aimed at categorizing and mapping the different types of land cover present in a given area [1–3]. Accurate LULC classification is essential for a wide range of applications, including urban planning, natural resource management, environmental monitoring, disaster assessment, and more [4, 5].

© The Author(s), under exclusive license to Springer Nature Switzerland AG 2024
K. K. Patel et al. (Eds.): icSoftComp 2023, CCIS 2031, pp. 79–93, 2024.
https://doi.org/10.1007/978-3-031-53728-8_7

Due to the following reasons, LULC classification is challenging:

- Spectral variability: The spectral signatures of land cover classes like Barren Land, Built-up, etc. can overlap, making it difficult to distinguish between them.
- Mixed pixels: Many pixels in a remote sensing image contain a mixture of different land cover classes.
- Radiometric noise: The spectral signatures of pixels can be affected by radiometric noise, which can further complicate the classification task [6].

In recent years, different classification algorithms and methods have been developed to extract meaningful information from remotely sensed data. Two commonly employed techniques for LULC classification are the Minimum Distance (MD) classifier and the Spectral Angle Mapper (SAM) classifier [7]. The MD classifier is a straightforward and intuitive method based on the concept of Euclidean distance. It involves calculating the distance between each pixel's spectral signature and the mean spectral signatures of predefined classes. The pixel is then assigned to the class with the minimum distance. MD is suitable for situations where class separability is well-defined and the data distribution resembles Gaussian distributions. However, it may struggle with handling illumination variations and non-linear spectral variations caused by atmospheric effects or other factors [8, 9]. The SAM classifier, on the other hand, is an angular-based method that measures the spectral similarity between pixels and reference spectra. SAM calculates the angle between the spectral vectors of a pixel and a reference spectrum, using the spectral angle as a measure of similarity. This approach makes SAM less sensitive to illumination variations and allows it to capture spectral shifts caused by environmental conditions. SAM is particularly effective when dealing with materials exhibiting different brightness levels while still maintaining similar spectral shapes. In a study compared the performance of MD, SAM, and a neural network classifier for LULC classification of Landsat 8 imagery [10]. In another study compared the performance of the MDC and SAM for LULC classification using Landsat ETM + imagery. The authors found that the SAM had a higher overall accuracy than the MDC [11]. The results showed that the SAM classifier outperformed the MD classifier and the neural network classifier in terms of overall accuracy and kappa coefficient This study aims to explore and compare the performance of the MD and SAM classifiers for LULC classification; By evaluating their accuracy, strengths, and limitations. We aim to provide insights into the suitability of each method for various types of landscapes and applications. The comparative analysis of these classifiers will contribute to better understanding their respective capabilities and guide practitioners in making informed decisions when selecting the most appropriate classifier for their specific LULC classification [2].

In the subsequent sections of this research, in second section information about study area is given, third section discuss about the methodology used which includes preprocessing of data, selecting training data and classification followed by accuracy assessment. The result and analysis are discussed in Fourth section and finally this research work is concluded in Fifth section. The outcomes of this study will aid in advancing in the field of remote sensing analysis and improving the accuracy of LULC mapping for diverse applications.

2   Area of Study

The area selected for the study is situated in the paithan taluka of the Aurangabad district in the Indian state of Maharashtra, which includes the Jaykwadi region and its surrounding areas which is shown in Fig. 1. It has the name of one of Asia's biggest earthen dams, the Jayakwadi Dam. The water from the dam is used to irrigate the region's rich agricultural area, which is well renowned for its fertility. The Jayakwadi Bird Sanctuary, a well-liked tourist site, is also located in the Jayakwadi area.

Landsat 8 Operational Land Imager satellite data, which was downloaded via the Earth Explorer data site (https://earthexplorer.usgs.gov/), is used. The Landsat 8 satellite image used in this study was captured on May 6, 2022, over Path 146 and Row 047, and processed using the Collection 2 (C02) processing method to Level 2 Surface Reflectance (L2SP). This image is in the GeoTIFF format and has the collection category T1 assigned to it. The scene ID is LC08_L2SP_146047_20220506_20220512_02_T1 for this image [12, 13].

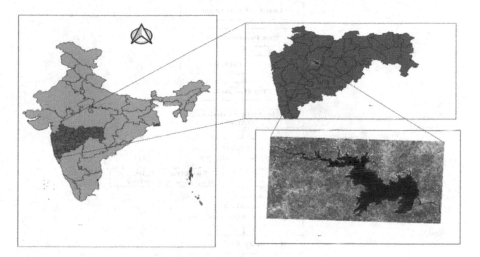

Fig. 1. Study Area Map

3   Methodology

Landsat 8 Level 2 data is used for the study which is already processed for atmospheric distortion, removing clouds and enhancing the spectral bands [14]; after those layers of image is been stacked, which allows the different features to be identified. Area of interest is then Clipped from the stacked image. This ensures that only the data is selected which is relevant to our study. Training data is selected from clipped image for the classification; These are training samples that have been manually selected and classified into their respective land cover classes. After training data is selected, classification is performed on the AOI. In this study two classification algorithms are used MD classifier and the SAM classifier. For Accuracy assessment a validation file is created using a high-resolution Google Earth image. This is a file that contains the ground truth data of land cover information for the area of Interest (AOI). Finally, accuracy is calculated for the classification; This can be done by comparing the classified pixels to the ground truth data. It is to draw conclusions from the study. This could involve discussing the strengths and weaknesses of the methods used, as well as the implications of the results. The Fig. 2. Shows the detailed methodology used in this research.

3.1 Preprocessing

The Level-2 image of Landsat- 8 underwent atmospheric and radiometric correction and was provided in the Geotiff format, ensuring accurate georeferencing and spatial information. The Landsat 8 Level-2 data provides surface reflectance values, making it suitable for quantitative analysis of land cover and environmental variables [13].

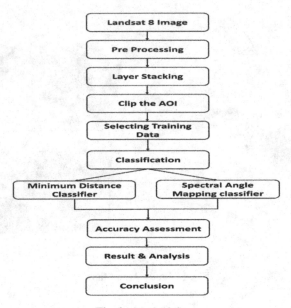

Fig. 2. Methodology

3.1.1 DN to TOA Reflectance Conversion

Data is given in Digital Numbers (DN), so it needs to be converted it into Top-of-Atmospheric Reflectance (TOA) values. It is a twostep procedure, first conversion of DN to radiance using Eq. 1, and later converting radiance to TOA using Eq. 2. For the Landsat 8 OLI/TIRS Sensor, TOA reflectance conversions were carried out using the

Dark object subtraction (DOS-1) atmospheric correction extension in the SCP Plugin in QGIS 3.28.8 [15, 16].

$$L_\lambda = mult_\lambda * DN + add_\lambda \qquad (1)$$

Where: L_λ is the cell value as radiance
DN is the cell value digital number
mult and add values listed in .MTL file

$$\rho\lambda = \pi * L\lambda * d^2/ESUN\lambda * cos\theta_s \qquad (2)$$

Where: $\rho\lambda$ = Unitless plantary reflectance
L_λ = spectral radiance (from earlier step)
d = Distance between Earth-Sun in astronomical units
$ESUN_\lambda$ = mean solar exoatmospheric irradiances
θs = solar zenith angle

3.1.2 Layer Stacking

Layer stacking is the process of combining multiple individual raster layers or bands into a single multiband image [17, 18]. These individual layers often represent different spectral or spatial information captured by various sensors or satellite passes. The resulting stacked image contains all the selected layers, allowing for more comprehensive analysis and interpretation of the data [9, 17, 19]. Layer stacking allows feature extraction because different bands or layers might highlight distinct features on the Earth's surface; Combining these layers can enhance the ability to identify specific land cover types, geological features, or other characteristics [18, 20]. Following Table 1 Shows Band list of Landsat 8 [13].

Layer Stacking allows multiple bands with different spectral information helps to improves the accuracy of land cover classification algorithms and it also helps in classification. More spectral information enables better differentiation between land cover classes. Arranging different bands to different order in such a way that it can provide more visually informative representations of the data. This can reveal hidden patterns or correlations in the information. Below Table 2 shows different band combinations for different feature extraction [9].

Table 1. List of Landsat 8 Spectral Bands

Band No	Channel (μm)	Spatial Resolution (m)
1	Coastal Aerosol (0.43–0.45)	30
2	Blue (0.45–0.51)	30
3	Green (0.53–0.59)	30
4	Red (0.64–0.67)	30
5	Near-Infrared (0.85–0.88)	30
6	SWIR-1(1.57–1.65)	30
7	SWIR-2 (2.11–2.29)	30
8	Panchromatic (0.50–0.68)	15
9	Cirrus (1.36–1.38)	30
10	TIRS-1 (10.6–11.19)	100
11	TIRS-2 (11.5–12.51)	100

Table 2. Different Band Combinations for Feature Identification [17, 18]

Band Composite	Band Combination	Features
Natural Color	4-3-2	This is similar to what human eyes can see
False Color (urban)	7-6-4	This combination shows vegetation in shades of green, urban areas have blue color and soils in different shades of brown
Color Infrared	5-4-3	This band combination is useful in analyzing vegetation because chlorophyll reflects near infrared light.Here areas with red have better vegetation health. Water is in dark and urban areas have white color
Agriculture	6-5-2	This is band combination is used for crop monitoring, healthy vegetation appears dark green and barren land has magenta hue color
Atmospheric Penetration	7-6-5	This combination takes advantage of the SWIR bands to penetrate atmospheric haze and improve visibility of surface features

(*continued*)

Table 2. (*continued*)

Band Composite	Band Combination	Features
Healthy Vegetation	5-6-2	Since healthy plants actively reflect near-infrared light in this combination, the near-infrared band (Band 5) is very helpful for highlighting vegetation that is in good health
Land and Water	5-6-4	In this combination, earthly regions seem brighter, whereas ocean bodies often appear dark or black. This is due to the reason that water absorbs more infrared and shortwave infrared light than land does, giving the image a clear contrast
Natural color With Atmospheric noise Removal	7-5-3	This combination can provide a clearer representation of the Earth's surface by minimizing the impact of atmospheric effects. It's particularly useful when trying to visualize land cover features without the interference of atmospheric scattering
Shortwave Infrared	7–5–4	This combination can be valuable for tasks such as identifying different soil types, mapping moisture variations, detecting mineral deposits, and monitoring changes in vegetation health
Vegetation health Analysis	6-5-4	This combination is particularly useful for tasks such as assessing vegetation health, land cover classification, monitoring agricultural areas, and detecting areas of stress or disturbance in vegetation

Layer stacking must take into account details like radiometric calibration and geometric alignment, as well as precise layer alignment to avoid distortions in the stacked image. Layer stacking and subsequent analysis are capabilities offered by remote sensing software programs for use in a variety of academic and practical applications [9].

3.1.3 Clip ROI

The Landsat 8 scene size is 185 km cross-track X 180 km along track, clip our area of interest from the large image; Because it can be time-consuming and computationally expensive to process them. Clipping the image to the extent of AOI can significantly reduce the size of image, making it easier and faster to process [21]. It also allows us to focus on specific region. Satellite images can often contain unwanted features, such as clouds, shadows, or snow. Clipping the image to our AOI can help to remove these unwanted features, making it easier to see the data that you are interested in [22]. In general, it is a good idea to clip an area of interest from a satellite image whenever we only need to analyze or focus on a specific region. This can save time, resources, and improve the accuracy analysis. Following Fig. 3(a) and 3(b) shows Landsat 8 image before clipping and after clipping.

Fig. 3. (a) Landsat 8 Image, (b) Clipped Area of Interest from Image

3.2 Selecting Training Samples

While selecting training data from clipped image for supervised classification, it is important to consider the LULC classes. This will help to determine how many training samples we need to collect for each class [23]. Different land cover classes have different spectral signatures, which means that they reflect light differently in different wavelengths. Selecting different band combinations as shown in Table 2. While selecting the training sample, consider spectral characteristics, spatial distribution and size of the training samples; the size of the training samples will affect the accuracy of classification. Larger training samples are generally more accurate than smaller training samples; The training data must include all important land-cover classes within the clipped image to avoid wrong allocations of pixels to classes of interest [9, 16, 23].

3.3 Classification

Classification is a method assigns label to each pixel in an image based on its spectral properties. The labels can represent different LULC classes, such as Fallow Land, vegetation, Barren Land, water and built-up areas. Classification can be used for different applications such as LULC Cover mapping, Natural resource management, Environmental Studies and mainly for Disaster management. It helps to monitor changes in land use over time to plan for future development and possible threat in future for disaster management. Classification is a complex process that requires careful consideration of the data, the objectives of the study, and the available methods. However, it is a powerful tool that can be used to gain valued insights from the Earth's surface and its dynamics. The availability of RS data, the complexity of the terrain, the choice of image bands, the algorithm used for classification, the analyst's knowledge with the study area, and the analyst's previous expertise with the classifiers algorithm used all have an impact on the classification process and outcomes [6]. In this study MD and SAM of supervised classification are used [22].

3.3.1 Minimum Distance Classification

MD classification is one of the supervised algorithms used for classification that is used to classify each pixel in an image into various different land cover classes. The algorithm works by first creating a training set of pixels that have been labeled with their correct LULC class. The mean spectral signature of each land cover class is then calculated from the training set. When a new pixel is encountered, the algorithm calculates the Euclidean distance between the pixel's spectral signature and the mean spectral signatures of all of the land cover classes. The pixel is then assigned to the land cover class with the Minimum Distance [24].

For each pixel in the image, Euclidean distance is calculated between its feature vector and the mean feature vectors of each class. The formula for Euclidean distance between two vectors X and Y is shown in Eq. 3:

$$distance = \sqrt{\sum_{i=1}^{n} (X[i] - Y[i])^2} \tag{3}$$

MD classification is a simple and relatively fast algorithm to implement. However, it can be sensitive to noise in the data and can produce misclassifications if the training set is not representative of the entire image; It can be used to classify pixels in images with a large number of land cover classes [9, 24]. Disadvantages of MD classification are, the training set used may have an impact on it. Misclassifications may result if the training data is not accurate, It can be computationally expensive for images with a large number of pixels. MD classification is a simple and effective algorithm for classifying pixels in an image [24].

3.3.2 SAM Classification

SAM allows the quick mapping of the spectral similarity of reference spectrum with test Spectrum. The mathematical formula for SAM to find the angles formed between the reference spectrum and the test spectrum treating them as vectors in a space with dimensionality equal to the number of bands [25]. SAM presents the following formulation: Eq. 4.

$$cos \ \alpha = \frac{\sum_{XY}}{\sqrt{\sum_{X^2} \sum_{Y^2}}} \tag{4}$$

Where: α = Angle between image spectrum and reference spectrum

X = Image spectrum
Y = Reference spectrum

Considering two bands as vectors in a space with n dimensions corresponding to the number of bands, A spectral classifier, the SAM approach, determines how similar two spectra are [26]. To categorize each pixel, a certain threshold is applied using the angular information between pixel spectra as shown in Fig. 4. The classification process ignores pixels whose angles are beyond the maximum angular threshold which is in radians. This implies that the match will be better the closer the angle. In SAM, reference spectra for area of interest may be retrieved, either from the satellite data or from the existing spectral libraries or from field measurements. SAM supervised classifier primarily depend on spectral pattern's formed, and the data need not follow a normal distribution. Additionally, solar illumination parameters have no impact on the SAM's measurement of the angle between two spectral vectors [16, 27]. The spectrum mixing issue is SAM's biggest drawback. The assumption that the endmembers used to categories the picture reflect the pure spectra of a reference material is the most false assumption made with SAM. Images with average spatial resolution seem to have this issue [26].

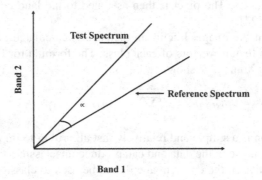

Fig. 4. For two band image Plot of Test Spectrum with Reference Spectrum [28]

In Fig. 5 shows classified image of MD and SAM classifier; MD classifier has classified the majority of area as barren land because of its spectral similarity and SAM classified it into its proper class as fallow land or vegetation.

3.4 Accuracy Assessment

Accuracy assessment allows us to evaluating the accuracy of map generated using the algorithm; for accuracy assessment, classification is compared to actual data to see how accurately the classification captures the actual data. The accuracy of result is assessed using a confusion matrix and the Kappa coefficient; calculated using Eq. 5 and Eq. 6 [29]. The confusion matrix was created by comparing the classified pixels to the reference data. The Kappa coefficient was calculated to measure the similarity between the classified and reference data.

$$Overall\ Accuracy = \frac{Number\ of\ correctly\ classified\ pixel\ (Diagnol)}{Total\ Number\ of\ Reference\ Pixels} \times 100 \quad (5)$$

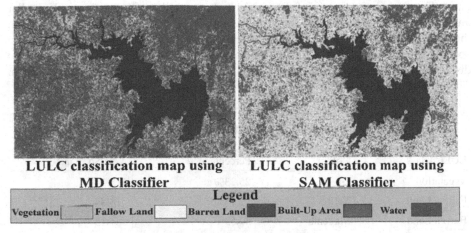

Fig. 5. LULC Classification Map

$$Kappa\ Coeficient\ =\ \frac{(TS \times TCS) - \sum(Column\ total \times Row\ Total)}{TS^2 - \sum(Column\ total \times Row\ Total)} \times 100 \quad (6)$$

Where: TS = Total Sample

TCS = Total Corrected Sample.

When studying image classification and LULC change detection, accuracy evaluation is crucial since it helps to comprehend and estimate the changes to a specific region properly [30]. High resolution google earth satellite image is used for ground truth validation. When compare the results with the validation file, SAM classifier provides a highest overall accuracy of 85.74% with 0.81 Kappa coefficient, whereas MD classifier provides an overall accuracy of 80.24% with 0.70 Kappa coefficient as shown in above Table 3.

Table 3. Kappa Coefficient and Overall Accuracy of classifier

Classifier Name	Overall Accuracy %	Kappa Coefficient
Minimum Distance	80.24%	0.70
Spectral Angle Mapper	85.74%	0.81

4 Result and Analysis

The SAM and the MD are both supervised classification algorithms used in this study. They both compare the spectral signature of each pixel with the spectral signatures of known classes of land cover. They approach it differently, though the SAM algorithm calculates the angle between the spectral signatures of the pixel and the land cover class. The angle is a measure of the similarity between the two signatures. A smaller angle indicates a closer match. The pixel is then classified to the land cover class with the smallest angle. The MD algorithm calculates the Euclidean distance between the spectral signatures of the pixel and the land cover class. The distance is a measure of the dissimilarity between the two signatures. A smaller distance indicates a closer match. The pixel is then classified to the land cover class with the smallest distance [9].

Table 4. Minimum Distance Classification Result

LULC Class	Colour	Percentage %	Area Sq. Km.
Vegetation	Green	25.23%	413.11
Fallow Land	Yellow	1.15%	18.84
Barren Land	Magenta	52.59%	861.13
Bulit Up	Red	2.48%	40.55
Water	Blue	18.55%	303.72

The Table 4 shows results of the MD Classifier (MDC) of a land cover image, MDC classification is a supervised classification algorithm that assigns appropriate class to each pixel in an image whose mean spectral signature is closest to the pixel's spectral signature. The table shows that Barren land cover type, covering 52.59% area of the image, Vegetation is second dominant land cover class which covers 25.23% of the area in image. Fallow land covers 1.15% of the image area, built up areas cover 2.48% of the image area, and water covers 18.55% of the image area.

Table 5. Spectral Angle Mapper Classification Result

LULC Class	Colour	Percentage %	Area Sq. Km.
Vegetation	Green	18.43%	301.83
Fallow Land	Yellow	39.83%	652.23
Barren Land	Magenta	17.48%	286.21
Bulit Up	Red	8.16%	133.55
Water	Blue	16.10%	263.54

The Table 5 shows the results of a SAM classification of a land cover image, SAM algorithm compare the spectral signature of each and every pixel in the image to a set of reference spectra. The reference spectra represent known land cover types, such as vegetation, barren land, fallow land, water, and built up, Table 5 shows that fallow land is the most dominant land cover type, covering 39.83% of total image area, whereas vegetation is second most dominant land cover type, covering 18.43% of the total image area. Barren land is the third most dominant land cover type, which covers 17.48% of the total image area. Built up areas cover 8.16% of the image area, and water covers 16.09% of the image area.

SAM has given more accurate result as compared to MD as shown in Table 5, because MD method faces the problem in classification of fallow land, barren land and built-up area. These classes have very much similar spectral signature and it get misclassified in other classes; especially barren land class is much more susceptible to wrong classification. As the MD classifier does not consider the direction of the spectral signature it can occasionally misclassified with the identical spectral signature. The SAM has given a highest accuracy as compared to Minimum Distance classifier, because it also considers the directionality of the spectral signature. Landsat 8 image selected for the study is of May month, in the month of May formers harvest the crops, so much of the land as fallow Land, so it is classified as barren land or Bult Up because of the spectral signature. MD classifier classified only 1.15% land as fallow land and 52.59% land as barren land, but in actual the majority of the land is used for farming and only few lands is barren. SAM classifier classified 39.83% land as fallow land and only 17.48% land as Barren Land, which is more accurate as compared to MD classifier.

5 Conclusion

Land Use and Land Cover classification gives the overall mapping of given area, it needs to be accurate which is crucial for wide range of applications as well as future planning. In this study Landsat-8/OLI level L2 image were used, which is atmospherically corrected data and "ready to use data" as there is no need to implement any corrections. In this study, the DN values are converted to TOA reflectance; because DN values are associated with impurities due to Atmospheric influences so they need to be converted. After preprocessing classification of LULC were performed, for that purpose five classes of LULC were considered. The highest overall accuracy of 80.24% with kappa coefficient of 0.70 is given by Minimum Distance classifier algorithm whereas SAM has given the overall accuracy of 85.74% with kappa coefficient 0.81.SAM has given a highest accuracy as compared to Minimum Distance classifier, because it also considers the directionality of the spectral signature. Since the direction it is not taken into consideration by the MD classifier it occasionally misclassified with the identical spectral signature but deferring directional pattern. For this study image was taken in the 6 may 2022 and in this duration, farmers harvest their crops, so much of the land is Fallow Land. SAM has accurately classified the land as follow land where as due to the similar spectacle signature MD has classified the land as a Barren Land. As compared to MD classifiers, SAM has provided the highest accuracy, and also SAM provides satisfactory value for each type of land class as compared to Minimum Distance classifier. Future research will extend by using Hyperspectral data so, it will be possible to go for further sub classification of data into distinct classes such as shallow water, deep water and pond.

Acknowledgement. The authors would like to thank the Mahatma Jyotiba Phule fellowship for providing financial assistance. The author would like to thanks Dr. G. Y. Pathrikar College of Computer Science and Information Technology, MGM University, Chhatrapati Sambhaji Nagar, Maharashtra, India.

References

1. Prakasam, C.: Land use and land cover change detection through remote sensing approach: a case study of Kodaikanal taluk, Tamil nadu. Int. J. Geomatics Geosci. **1**(2), 150 (2010)
2. Nagne, A.D., Gawali, B.W.: Transportation network analysis by using Remote Sensing and GIS a review. Int. J. Eng. Res. Appl. **3**(3), 70–76 (2013)
3. Tiwari, A., Karwariya, S.K., Tripathi, S.: Monitoring land use/cover change using digital classification techniques: a case study of Sadhera Mines, Satna, Madhya Pradesh, India. Eur. J. Eng. Technol. Res. **1**(1), 34–38 (2016)
4. Yao, X., et al.: Land use classification of the deep convolutional neural network method reducing the loss of spatial features. Sensors **19**(12), 2792 (2019)
5. Wambugu, N., et al.: A hybrid deep convolutional neural network for accurate land cover classification. Int. J. Appl. Earth Obs. Geoinf. **103**, 102515 (2021)
6. Al-Doski, J., Mansorl, S.B., Shafri, H.Z.M.: Image Classification in Remote Sensing. Department of Civil Engineering, Univ. Putra, Malaysia, vol. 3, no. 10 (2013)
7. Mather, P., Tso, B.: Classification Methods for Remotely Sensed Data. CRC Press, Boca Raton (2016)
8. Vibhute, A.D., Gawali, B.W.: Analysis and modeling of agricultural land use using remote sensing and geographic information system: a review. Int. J. Eng. Res. Appl. **3**(3), 81–91 (2013)
9. Lillesand, T., Kiefer, R.W., Chipman, J.: Remote Sensing and Image Interpretation. Wiley, Hoboken (2015)
10. Aung, M.S.H., Khaing, Z.Z., Thu, K.M., Zin, T.T.: A comparison of pixel-based classification algorithms for land use/land cover classification using Landsat 8 imagery. Earth Syst. Environ. **6**(1), 621–630 (2022)
11. Ozturk, T.H., Avci, M., Coskun: A comparison of spectral angle mapper (SAM) and minimum distance classifier (MDC) methods for land use/land cover classification. ISPRS J. Photogramm. Remote Sens. **83**, 122–132 (2013)
12. Kandekar, V.U., et al.: Surface water dynamics analysis based on sentinel imagery and Google Earth Engine Platform: a case study of Jayakwadi dam. Sustain. Water Resour. Manag. **7**(3), 44 (2021)
13. Landsat, E.: Landsat 8-9Operational Land Imager (OLI)-Thermal Infrared Sensor (TIRS) Collection 2 Level 2 (L2) Data Format Control Book (DFCB), United States Geol. Surv. Reston, VA, USA (2020)
14. Vermote, E., Justice, C., Claverie, M., Franch, B.: Preliminary analysis of the performance of the Landsat 8/OLI land surface reflectance product. Remote Sens. Environ. **185**, 46–56 (2016)
15. Congedo, L.: Semi-Automatic Classification Plugin: A Python tool for the download and processing of remote sensing images in QGIS. J. Open Source Softw. **6**(64), 3172 (2021)
16. Lin, S.-K.: Introduction to Remote Sensing. By James B. Campbell and Randolph H. Wynne, The Guilford Press (2011). 662 pages. Price:£80.75. ISBN 978-1-60918-176-5. Molecular Diversity Preservation International (MDPI) (2013)
17. Kevin_butler and Kevin_butler: Band Combinations for Landsat 8. ArcGIS Blog (2019). https://www.esri.com/arcgis-blog/products/product/imagery/band-combinations-for-landsat-8/
18. GISGeography: Landsat 8 Bands and band Combinations. GIS Geogr. (2022). https://gisgeography.com/landsat-8-bands-combinations/
19. Ratnaparkhi, N.S., Nagne, A.D., Gawali, B.: Analysis of land use/land cover changes using remote sensing and GIS techniques in Parbhani City, Maharashtra, India. Int. J. Adv. Remote Sens. GIS **5**(1), 1702–1708 (2016). https://doi.org/10.23953/cloud.ijarsg.54

20. Ratnaparkhi, N.S., Nagne, A.D., Gawali, B.: A land use land cover classification system using remote sensing data. Changes **7**(8) (2014)
21. Irons, J.R., Dwyer, J.L., Barsi, J.A.: The next Landsat satellite: the Landsat data continuity mission. Remote Sens. Environ. **122**, 11–21 (2012)
22. Nagne, A.D., Dhumal, R.K., Vibhute, A.D., Gaikwad, S., Kale, K., Mehrotra, S.: Land use land cover change detection by different supervised classifiers on LISS-III temporal datasets. In: 2017 1st International Conference on Intelligent Systems and Information Management (ICISIM), pp. 68–71 (2017)
23. Van Niel, T.G., McVicar, T.R., Datt, B.: On the relationship between training sample size and data dimensionality: Monte Carlo analysis of broadband multi-temporal classification. Remote Sens. Environ. **98**(4), 468–480 (2005)
24. Nagne, A.D., Vibhute, A.D., Dhumal, R.K., Kale, K.V., Mehrotra, S.C.: Urban LULC change detection and mapping spatial variations of Aurangabad city using IRS LISS-III temporal datasets and supervised classification approach. In: Data Analytics and Learning: Proceedings of DAL 2018, pp. 369–386 (2019)
25. De Carvalho, O.A., Meneses, P.R.: Spectral correlation mapper (SCM): an improvement on the spectral angle mapper (SAM). In: Summaries of the 9th JPL Airborne Earth Science Workshop, p. 2. JPL Publication 00-18 (2000)
26. Girouard, G., Bannari, A., El Harti, A., Desrochers, A.: Validated spectral angle mapper algorithm for geological mapping: comparative study between QuickBird and Landsat-TM. In < XXth ISPRS Congress, Geo-Imagery Bridging Continents, Istanbul, Turkey, p. 23 (2004)
27. Hasan, E., Fagin, T., El Alfy, Z., Hong, Y.: Spectral Angle Mapper and aeromagnetic data integration for gold-associated alteration zone mapping: a case study for the Central Eastern Desert Egypt. Int. J. Remote Sens. **37**(8), 1762–1776 (2016)
28. Kruse, F.A., et al.: The spectral image processing system (SIPS)—interactive visualization and analysis of imaging spectrometer data. Remote Sens. Environ. **44**(2–3), 145–163 (1993)
29. Islami, F.A., Tarigan, S.D., Wahjunie, E.D., Dasanto, B.D.: Accuracy assessment of land use change analysis using Google Earth in Sadar Watershed Mojokerto Regency. IOP Conf. Ser. Earth Environ. Sci. **950**(1) (2022). https://doi.org/10.1088/1755-1315/950/1/012091
30. Nagne, A.D., et al.: Performance evaluation of urban areas land use classification from hyperspectral data by using mahalanobis classifier. In: 2017 11th International Conference on Intelligent Systems and Control (ISCO), pp. 388–392 (2017)

A U-Net Based Approach for High-Accuracy Land Use Land Cover Classification in Hyperspectral Remote Sensing

Atiya Khan[1,3] , Chandrashekhar H. Patil[1(✉)] , Amol D. Vibhute[2] ,
and Shankar Mali[1]

[1] School of Computer Science, Dr. Vishwanath Karad MIT World Peace University, Pune, MH,
India
chpatil.mca@gmail.com
[2] Symbiosis Institute of Computer Studies and Research (SICSR), Symbiosis International
(Deemed University), Pune 411016, MH, India
[3] G H Raisoni College of Engineering, Nagpur, MH, India

Abstract. Deep learning has been demonstrated to have significant potential in the classification of hyperspectral images (HSI). Hyperspectral imaging has gained more recognition in recent years in the area of computer vision research. Its potential to handle remote sensing-related issues, particularly those related to the agricultural sector, has led to its rising popularity. Due to the significant spectral band redundancy, the small number of training samples, and the non-linear relationship between spatial position and spectral bands, HSI classification is a challenging task. Therefore, we propose machine and deep learning-based models to classify the land features with the highest accuracy. Effective bands have been discovered by applying principal component analysis (PCA) to minimize the dimensionality of hyperspectral images. In this work, we evaluate the land use and land cover (LULC) classification efficiency of three different algorithms like Support Vector Machine (SVM), Spectral Angle Mapper (SAM) and U-Convolutional Neural Network (U-net).Using Hyperion images, we demonstrate and evaluate the findings from each method. In this work, we apply deep convolutional neural networks for the classification of high-quality remote sensing images. Semantic image segmentation is used for U-Net frameworks. In the Nagpur district, we map the existence or lack of vegetation and agricultural land using the U-net neural network architecture for Hyperion images.

Keywords: Deep Learning · vegetation classification · Convolution Neural Network · hyperspectral imagery · U-net

1 Introduction

In recent years, researchers have paid a lot of attention to hyperspectral imaging systems. These systems' sensors allow more than a hundred spectral wavebands to be simultaneously captured for every pixel in an image. This comprehensive spectrum data raises the

K. K. Patel et al. (Eds.): icSoftComp 2023, CCIS 2031, pp. 94–106, 2024.
https://doi.org/10.1007/978-3-031-53728-8_8

prospect of more precisely identifying objects, materials, or areas of interest. Restoring images and extracting data are the goals of satellite image processing. The HSI technology must primarily address many hurdles it still faces. For example, hyperspectral images acquired by satellite or aircraft imaging spectrometers can include hundreds of dimensions in their spectrum reflectance values [1]. The composition and spatial distribution of ground features, atmospheric conditions, sensors, and the surrounding environment all have an impact on the spatial variability of spectral data. Remote sensing images [2] taken at different times capture geographic information about the dynamic changes that occur in a place, especially the formation of urban, lake, and other ecosystems. This has important ramifications for science. Deep learning has recently gained popularity, opening up new opportunities for analyzing massive data sets. The deep network's effective learning capability has enhanced the initial image processing approaches for reducing dimensionality, denoising, and other issues. Deep neural networks (DNNs) are an efficient approach for feature extraction and classification. Choosing appropriate spectral-spatial kernel [3] features is difficult because of noise and band correlation. In most cases, convolutional neural networks (CNNs) with fixed-size receptive fields are used to address this. However, when both backward and forward propagations are utilized to optimize the network, neurons lack the ability to successfully modify its dimensions and cross-channel dependencies. The DNNs [4] are efficient for processing images that can also be employed for other applications. The HOADTL-CC model is used to recognize and categorize crops on hyperspectral remote-sensing images.

Reconstruction and classification are carried out simultaneously by a generator in the Multitask generative adversarial network (MTGAN) [5]. Using an encoder and decoder sub-network, the reconstruction objective aims to reconstruct the input HSI cube, whereas the classification target aims to identify the input cube's category using a CNN. Multiple features are retrieved initially, followed by multiple CNN blocks [6]. 2D-CNN simply performs convolution on the spatial dimension for HSI classification; however, we prefer to collect spectral information stored in several frequency bands as well as spatial information. The 3D-CNN kernel [7] is capable of extracting band and spatial representations of features from HSI, albeit at the expense of greater computational complexity and a sluggish model. The faster selective kernel mechanism network (FSKNet) was built in response to this. The 3D-CNN is used to extract all features while lowering both spatial and spectral dimensionality. An efficient minority class oversampling strategy for HSI data classification problems involving class imbalance is the 3D-HyperGAMO generative adversarial minority oversampling method [8]. The system consists of a 3D-generator network that uses intermediate models for a noise vector generated by the conditioned features mapping unit and 3D-HSI patches to create fresh samples of a given class. A 3-D classification network that is also utilised to understand the details of each class is used to classify the samples, both fake and authentic, into the appropriate groups. Using unlabeled input to enhance CNN generalization and discrimination in the setting of generative adversarial networks (GANs) [9]. CNNs have a smaller impact on the over fitting issue. In particular, two frameworks are developed: the 1D-GAN, which uses spectral vectors, and the 3D-GAN, which integrates spectral and spatial features [10]. In contrast to other innovative approaches, these two architectures performed exceptionally well in the extraction of features and classification of images.

The deep CNN outperforms the regular CNN in terms of classification accuracy due to the use of GANs.

Further challenges arise when using the U-Net architecture [11] with earth observation data, such as selecting the optimum algorithms for kernel initialization, loss optimizer, learning rate activation, and loss. The depth include number of layers, kernel size, and starting filter count can all be changed while constructing a U-Net design. The ability of a model to accurately acquire information from the training datasets is greatly influenced by each of these factors. A diminishing gradient problem that leads to insufficient learning can arise when CNNs employ activation functions with a narrow range of values, such as sigmoid activation functions. These values were used to train an optimized U-Net model after the best parameters had been determined.

The original U-Net parameters were also employed to train a model that could be compared to the upgraded version [12]. More convolutional filters and larger kernel sizes resulted in more accurate models. As a result of our findings, we recommend using 56 convolutional filters with 5×5 kernel sizes for earth-observing LULC applications on this scale while retaining the U-Net initial depth of five layers.

The precise objectives of this research work were to:

- To improve LULC classification accuracy by creating synthetic hyperspectral data which increases the variety and depth of the training dataset. This contributes to tackling the difficulties of data labelled scarcity and intra-class heterogeneity in agricultural contexts.
- To improve the U-Net architecture for classifying LULC by utilizing lightweight depth-wise separable convolutions and residual connection, to improve accuracy and reduce model complexity.
- To classify and compare the LULC using SVM, SAM, and U-net classifiers.

This study is divided into four sections. The first of which is an introduction providing baseline knowledge on hyperspectral remote sensing. The study area and datasets used are explained in Sect. 2. The methodologies for pre-processing, data dimensionality reduction, feature extraction, and classification are provided in Sect. 3. A description of the estimated output and the results are included in Sect. 4. Finally, in the last section, conclusions with future implications are summarized.

2 Study Area and Datasets

The present study was carried out for the Winter/Rabi crop growing season in Nagpur district from October 2022 to February 2023. Figure 1 illustrates the location of the Nagpur district in Maharashtra state, India, between latitudes of 21.146633 N and longitudes of 79.088860 E. The area of the district is around 9897 square kilometers. The town is situated between 274.5 and 652.70 m above sea level and has a 28 per cent forest cover. In the Nagpur district, all arable land is divided into three types: rice fields, dry crop fields, and watered or garden fields. Dry agricultural fields are further divided into rabi (late monsoon) and kharif (early monsoon) because they rely on the monsoon. While Nagpur's dry season is normally clear and warm all year long, the rainy season is dim and cloudy. The average annual temperature is between 14 °C and 43 °C, with lows of 10 °C and highs of 45 °C being extremely rare.

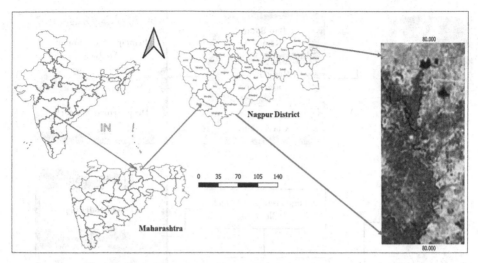

Fig. 1. Study area map and its satellite image (RGB composite).

The present study uses EO-1 satellite Hyperion data that captures 242 distinct spectral channels with 10-nm bandwidth spanning from 0.357 to 2.576 μm. The Level 1R collection of data includes metadata in binary and ASCII formats as well as radiometrically corrected images in HDF format. A Hyperion level one GST product including radiometric, geometric, and systematic terrain corrections was downloaded and used for this investigation. HDF file formats and Geographic Tagged Image-File Format were used to gather Hyperion imagery. For data processing level 0 and level 1 products were used.

3   The Proposed Methodology

The proposed approach is presented in Fig. 2 which focuses on EO-1 Hyperion image preprocessing, reducing the dimensionality, extracting the selective features and executing the SVM, SAM, and U-net models to classify the land features and map accordingly.

3.1   Input Data Preprocessing

The pre-processing of the hyperspectral image is essential to remove the image distortions, renovate the images, and eliminate the unwanted bands. In addition, radiometric, atmospheric corrections and geometric corrections are also needed for further processing. The original Hyperion image downloaded on 29, November 2022 has 242 spectral bands ranging from 400–2500 nm spectral region with 10 nm bandwidth. However, some bands were uncalibrated that needed to be deleted before applying further algorithms [15]. A total of 196 bands had accurate calibration, while the other 62 and 24 were referred to as noisy bands because they contained no data. In EO-1 Hyperion data, the bands from 1–7, 58–76, and 221–242 were automatically adjusted to 0 by the data supplier [13]. The SWIR sensor has 172 bands, while the Hyperion VNIR sensor has 70 bands [28].

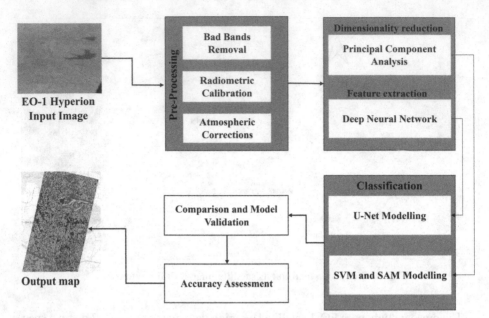

Fig. 2. The flowchart of the proposed approach

Additionally, bands 167–180 and 121–126 should be ignored during processing since they have a lot of noise due to severe water vapor absorption [29].

3.1.1 Radiometric Calibration

To build the radiance graph for each of the image classes, radiometric calibration is necessary to produce the spectrum profile in the required parameters, which are expressed in terms of radiometric values. The scanned image will be recorded as a signed integer known as a digital number (DN) [17]. The DN must be transformed into radiance using the sensor's offset and gain. The detected radiance value can be used to adjust the image for atmospheric effects. The radiance spectra of the vegetation class are displayed in Fig. 3.

The pixel DN value will be converted to at-sensor radiance using scale factors. The scale factor for the VNIR composite raster is 40, and the scale factor for the SWIR composite is 80. Equation (1) can be used to solve the conversion of DN to radiance value.

$$\frac{VNIR}{40} \text{ and } \frac{SWIR}{80} \tag{1}$$

The top of the atmosphere (TOA) radiance is the amount of radiation that was registered by a sensor during reflection events. Radiometric calibration is the procedure for obtaining TOA radiance values for DNS. The sensor gains and bias in each spectral band should be calibrated before obtaining the TOA radiance values for DNs using Eq. (2).

$$\text{Radiance } L\lambda \, (DN * Gain) + Bias \tag{2}$$

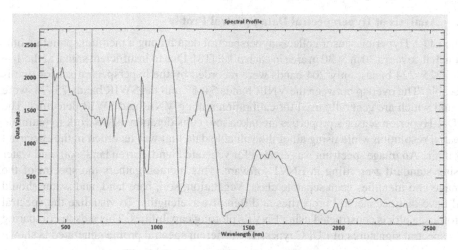

Fig. 3. Radiance profile of vegetation class

Where Bias λ and Gain λ are the bias and gain values for each spectral band (λ), respectively.

To calibrate the data to TOA reflectance, ENVI employs the reflectance gains and offsets from the input file's metadata. When reflectance gains and offsets are missing from the input file, ENVI uses Eq. (3) to compute TOA reflectance.

$$\rho_\lambda = \frac{\pi L\lambda d2}{ESUN\lambda \sin \theta} \tag{3}$$

Where: L_λ = Radiance in units of W/(m2 * sr * μm), d = Earth-sun distance in astronomical units, $ESUN_\lambda$ = is the mean solar irradiance for each spectral band, 0 = Sun elevation in degrees.

3.1.2 Atmospheric Corrections

The atmospheric correction of hyperspectral images is a significant task. The atmospheric retrieval components are immediately completed using the image's data, and it works well without any further information. VNIR, SWIR and 3000 nm spectral ranges are all corrected using the principles of FLAASH atmospheric correction model [14]. Based on measurements of TOA radiance or TOA reflectance, surface reflectance values are calculated using atmospheric adjustment techniques. The atmospheric correction has been done using the FLASH model given in ENVI software [13]. A radioactive transfer code is used to construct look-up tables for the FLAASH [14] atmospheric correction project. After removing the bad bands, bad columns, and de-striping, the resized 195 bands from the Hyperion dataset were corrected for atmospheric errors using ENVI's atmospheric correction model's FLASH model. The FLASH produced better outcomes concerning of the spectral profile achieved at the end of this stage.

3.2 Analysis of Hyperspectral Data Spectral Profile

The EO-1 Hyperion sensor collects hyperspectral data having a medium spatial resolution that covers a 30 m × 30 m area in each pixel [18]. Due to insufficient signal at the 1–9 and 225–224 bands, only 204 bands were recorded by the hyperspectral sensor that is useable. The overlap between the VNIR bands 54–57 and the SWIR bands 75–78, were found which are generally used for calibration among VNIR and SWIR detectors. The EO-1 Hyperion sensor's properties are taken into consideration as we work to maintain spatial resolution while using all of the valuable data that was recorded in the spectrum domain. An image spectrum was created for vegetated land, barren land, soil, and water using standard z-profiling in ENVI software. This method gathers the spectra of the image and identifies each separate class. Vegetation, soil, bare land, and water should all have distinct spectral properties at different wavelengths. To visualize the spectral profiles, a plot is constructed using ENVI's plotting capabilities. This aids in comparing the spectral signatures of LULC types. The different spectral profile generated is shown in Fig. 4.

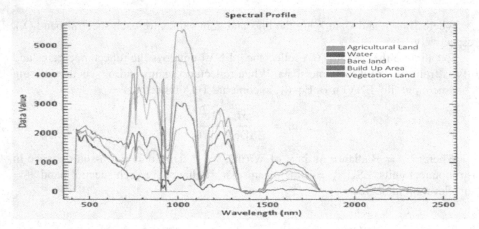

Fig. 4. Spectral response of various land cover types

3.3 Classification Algorithms

3.3.1 Support Vector Machine

The SVM is another popular supervised classification method [15] used in the present study that maximizes the distance among classes by dividing them into n-dimensional spaces with the least degree of misclassification through the creation of a hyperplane. It was found that, although SVM is typically used in conjunction with the radial basis function kernel, the linear kernel outperformed it in this particular set of data during classification. In order to improve the accuracy of land use classification and maximize the use of satellite imagery in the performance of LULC classification, the SVM approach is combined using texture and spectral data [19] to obtain information about the land-use mining zone.

3.3.2 Spectral Angel Mapper

The SAM method is an auto-generated supervised spectral classification machine learning technique that uses the angle between the spectra to calculate the spectral resemblance among the image's spectra and reference spectra in an n-dimensional space (where n is the spectral band number). As a result, SAM [22] only use angle information to determine the pixel spectra. The angle difference between the reference and image spectra is represented as a level of similarity, where a high angle denotes low similarity and a narrow-angle indicates high similarity.

3.3.3 U-Net Convolutional Neural Network

Based on convolutional filters, the U-net design reduces the image quality to different resolutions in order to identify patterns and texture in various sizes [20]. Every level of the U-net corresponds to a degradation phase, including filtering at that level. The links between the images are kept scale-appropriate by up scaling the coarsest scales close to the original resolution. The term "U-net" has been assigned to this concept because to the layering structure's schematic shape [21]. It was used to categorize distinct spots on microscope histology slides.

To map different types of LULC, the U-Net is used in this study work as a supervised semantically segmented network. The U-Net has been implemented using ENVI's deep learning software. Downscaling [23], which increases resilience against imagery distortion, and upscaling, which restores object features and decodes information about the input imagery, make up the two primary parts of the U-Net design.

3.4 Training Datasets

The models have to be trained across multiple class categories to use the classifiers. The SVM, SAM, and U-net are trained on discrete imagery patches. Even though our tests showed that sizes between 128 and 282 perform equally well, we chose to use a patch size of 128 by 128 pixels. Polygons in the shape file were used to save the patches. There are all three bands from the original Earth image in every patch. There is also a label image for each patch, in which each pixel has been annotated for the system's training. The geographic distribution of the patches is shown in Fig. 5, along with how the stratified sampling technique aided in achieving a more equitable distribution between classes of bigger and smaller areas.

To train the TensorFlow model, a large number of data have been selected from each class category: water, buildup, bare land, vegetable land, and agriculture land. Considering the simplicity of identifying these classes in the RGB images, they were selected. A theme map or class activation is produced by the TensorFlow model during training using the spectral or spatial data in the input images. Georeferenced activation and classification maps are produced by the model once its dataset training is complete. Fractional maps are created for each class category to improve the accuracy of these classification maps.

Fig. 5. Geographically located patches of the study area

4   Result and Discussions

During the image pre-processing, the marked field borders and the entire ROI are integrated. The only "real" space-borne hyperspectral sensor contains 242 spectral bands and a spatial resolution of 30 m. It gathers spectral information about things on the surface of the Earth. For further processing, the 167 calibrated bands are employed. FLAASH atmospheric correction has been done, which has altered the image's apparent reflectance by decreasing it in the blue and red regions and increasing it in the NIR and SWIR regions. The data are reduced in dimension and used as standard bands by the application of principal component analysis. The classification of the land cover has been accomplished, including built-up areas, agricultural fields, barren land, water, and vegetation. To produce images that are cropped with boundaries within the ROI, the ROI cropped mask images with selected indices are vectorized and the edges are smoothed.

To identify the pixels that in the clipped mask image must have the greatest degree of broad accuracy in the validation range, fold cross-validation is used to generate the train, test, and validation data.

At the first stage of classifying land cover, LULC shows great accuracy in segmentation results [24]. The validity of the pixels and the images is then verified visually by comparing the ground cover data that was generated from the vector data matrix. It shows the improved U-Net model. The network is made up of an encoding, a decoding, and a skip link [25]. For the encoding and decoding portions of each of the five layers, two feature extraction modules have been introduced. To restore the picture size and ensure that the fusion's resolution is the same across all layers, the basic feature map is treated to 2×2 up-sampling before each one. The last layer's 1×1 convolution identifies each pixel.

In order to increase classification accuracy [27], U-Net reduces network complexity, incorporates residual connections to obtain deep semantic information, and extracts multi-scale characteristics from hyperspectral images. Particularly in challenging tasks like the classification of hyperspectral images U-net deep learning models outperform traditional methods in terms of accurate classification as depicted in Fig. 6. In situations where traditional approaches can find it difficult to distinguish between classes with comparable spectral qualities, U-net's ability to collect precise features permits more accurate classification. We provide a novel feature extraction module that deviates from standard convolution in the U-Net architecture by using depth-wise separable convolution. This change effectively reduces network complexity while capturing features from hyperspectral images and deeper into semantic data, for improved accuracy. The overall accuracy of the U-Net model is more than the SAM and SVM methods. The experimental results are satisfactory for all the classification methods.

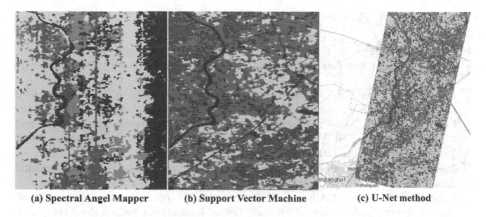

    (a) Spectral Angel Mapper   **(b) Support Vector Machine**   **(c) U-Net method**

Fig. 6. Hyperion image classification map using SAM, SVM and U-Net Model.

5 Conclusions

The Hyperion hyperspectral dataset, which consists of 242 bands, provides a useful method for accurately identifying a variety of patterns on the Earth's surface. The main difficulties in material identification are caused by the atmospheric conditions and the complex patterns on the surface of the Earth. The U-net model provides a method to take advantage of the visual potential present to distinguish between vegetated and bare terrain over vast geographical areas with great accuracy. U-net, a deep learning methodology, frequently produces superior accuracy than conventional spectral-based techniques like SAM and SVM. When properly trained and given precise reference spectra, SAM is successful, although it struggles with complicated spectral changes within classes. SVM, which is primarily intended for supervised learning, can achieve excellent accuracy when given well-labeled training data, but it is not frequently used for the categorization of unsupervised classification. In conclusion, U-net attains the maximum accuracy due to its capacity to understand complex characteristics and patterns.

Although U-Net is a deep learning model capable of extracting relevant characteristics from the data, future research activities may combine U-Net with methods for automatic extraction of features from hyperspectral data, whereby the understanding and classification performance of the model may be enhanced. To reliably evaluate the performance of various models, it will be necessary to create and maintain consistent benchmark datasets and assessment criteria tailored to hyperspectral LULC classification in future research work.

The preference among U-Net and other architecture, like ResNet, can be demonstrated by using the specific dataset for study. Depending on the requirements of the LULC classification task and the type of hyperspectral data, U-Net may perform more precisely and accurately than alternative approaches. A U-Net-based approach is usually used while LULC classification is carried out using remote sending images.

References

1. Aletti, G., Benfenati, A., Naldi, G.: A semi-supervised reduced-space method for hyperspectral imaging segmentation. J. Imaging **7**, 267 (2021). https://doi.org/10.3390/jimaging7120267
2. Gao, H., et al.: A hyperspectral image classification method based on multi-discriminator generative adversarial networks. Sensors **19**, 3269 (2019). https://doi.org/10.3390/s19153269
3. Roy, S.K., Manna, S., Song, T., Bruzzone, L.: Attention-based adaptive spectral–spatial kernel ResNet for hyperspectral image classification. IEEE Trans. Geosci. Remote. Sens. **59**(9), 7831–7843 (2021). https://doi.org/10.1109/TGRS.2020.3043267
4. Duhayyim, M.A., et al.: Automated deep learning driven crop classification on hyperspectral remote sensing images. Comput. Mater. Contin. **74**(2), 3167–3181 (2023)
5. Hang, R., Zhou, F., Liu, Q., Ghamisi, P.: Classification of hyperspectral images via multitask generative adversarial networks. IEEE Trans. Geosci. Remote Sens. **59**(2), 1424–1436 (2021). https://doi.org/10.1109/TGRS.2020.3003341
6. Agilandeeswari, L., Prabukumar, M., Radhesyam, V., Phaneendra, K.L.N.B., Farhan, A.: Crop classification for agricultural applications in hyperspectral remote sensing images. Appl. Sci. **12**(3) (2022). https://doi.org/10.3390/app12031670
7. Li, G., Zhang, C.: Faster hyperspectral image classification based on selective kernel mechanism using deep convolutional networks. Image Video Process. (eess.IV). https://doi.org/10.48550/arXiv.2202.06458
8. Roy, S.K., Haut, J.M., Paoletti, M.E., Dubey, S.R., Plaza, A.: Generative adversarial minority oversampling for spectral–spatial hyperspectral image classification. IEEE Trans. Geosci. Remote. Sens. **60**, 1–15, 5500615 (2022). https://doi.org/10.1109/TGRS.2021.3052048
9. Bai, J., Lu, J., Xiao, Z., Chen, Z., Jiao, L.: Generative adversarial networks based on transformer encoder and convolution block for hyperspectral image classification. Remote Sens. **14**, 3426 (2022). https://doi.org/10.3390/rs14143426
10. Zhu, L., Chen, Y., Ghamisi, P., Benediktsson, J.A.: Generative adversarial networks for hyperspectral image classification. IEEE Trans. Geosci. Remote Sens. **56**(9), 5046–5063 (2018). https://doi.org/10.1109/TGRS.2018.2805286
11. Sun, Y., Tian, Y., Xu, Y.: Problems of encoder-decoder frameworks for high-resolution remote sensing image segmentation: structural stereotype and insufficient. Neurocomputing (2018). https://doi.org/10.1016/j.neucom.2018.11.051
12. Clark, A., Phinn, S., Scarth, P., Clark, A.: Optimised U-Net for land use – land cover classification using aerial photography. PFG J. Photogramm. Remote Sens. Geoinf. Sci. **91**(2), 125–147 (2023). https://doi.org/10.1007/s41064-023-00233-3

13. Xi, B., Li, J., Diao, Y., Li, Y.: DGSSC: a deep generative spectral-spatial classifier for imbalanced hyperspectral imagery. IEEE Trans. Circuits Syst. Video Technol. **PP**, 1 (2022). https://doi.org/10.1109/TCSVT.2022.3215513

14. Vibhute, A.D., Kale, K.V., Dhumal, R.K., Mehrotra, S.C.: Hyperspectral imaging data atmospheric correction challenges and solutions using QUAC and FLAASH algorithms. In: 2015 International Conference on Man and Machine Interfacing (MAMI), Bhubaneswar, India, pp. 1–6 (2015). https://doi.org/10.1109/MAMI.2015.7456604

15. Aneece, I., Thenkabail, P.S.: Classifying crop types using two generations of hyperspectral sensors (hyperion and DESIS) with machine learning on the cloud. Remote Sens. **13**, 4704 (2021). https://doi.org/10.3390/rs13224704

16. Singh, D., Singh, R.: Evaluation of EO-1 hyperion data for crop studies in part of indo-Gangatic plains: a case study of Meerut District. Adv. Remote Sens. **4**, 263–269 (2015). https://doi.org/10.4236/ars.2015.44021

17. Kaushik, M., Nishan, R., Jayanth, R., Rao, K., Prasantha, H.S.: Pre-processing of E0–1 hyperion data. Int. J. Eng. Dev. Res. (IJEDR) **10**(6), f469–f476 (2022). https://ijcrt.org/papers/IJCRT22A6670.pdf. ISSN 2321-9939

18. Teodor, C., Alzenk, B., Constantinescu, R., Datcu, M.: Unsupervised classification of EO-1 hyperion hyperspectral data using Latent Dirichlet Allocation. In: International Symposium on Signals, Circuits and Systems ISSCS2013, Iasi, Romania, pp. 1–4 (2013). https://doi.org/10.1109/ISSCS.2013.6651211

19. Wang, L., Jia, Y., Yao, Y., Xu, D.: Accuracy assessment of land use classification using support vector machine and neural network for coal mining area of Hegang City, China. Nat. Environ. Pollut. Technol. **18**, 335–341 (2019). e-ISSN 2395-3454

20. Flood, N., Watson, F., Collett, L.: Using a U-net convolutional neural network to map woody vegetation extent from high resolution satellite imagery across Queensland, Australia. Int. J. Appl. Earth Obs. Geoinf. **82**, 101897 (2019). https://doi.org/10.1016/j.jag.2019.101897

21. Ronneberger, O., Fischer, P., Brox, T.: U-net: convolutional networks for biomedical image segmentation. In: Navab, N., Hornegger, J., Wells, W.M., Frangi, A.F. (eds.) MICCAI 2015. LNCS, vol. 9351, pp. 234–241. Springer, Cham (2015). https://doi.org/10.1007/978-3-319-24574-4_28

22. Khan, A., Vibhute, A.D., Mali, S., Patil, C.H.: A systematic review on hyperspectral imaging technology with a machine and deep learning methodology for agricultural applications. Ecol. Inform. **69**, 101678 (2022)

23. Singh, G., Singh, S., Sethi, G., Sood, V.: Deep learning in the mapping of agricultural land use using sentinel-2 satellite data. Geographies **2**, 691–700 (2022). https://doi.org/10.3390/geographies2040042

24. Kumar, M.S., Jayagopal, P.: Ecological Informatics Delineation of field boundary from multispectral satellite images through U-Net segmentation and template matching. Ecol. Inform. **64**, 101370 (2021). https://doi.org/10.1016/j.ecoinf.2021.101370

25. Yu, H., Jiang, D., Peng, X., Zhang, Y.: A vegetation classification method based on improved dual-way branch feature fusion U-net. Front. Plant Sci. **13**, 1047091 (2022). https://doi.org/10.3389/fpls.2022.1047091

26. Vibhute, A.D., Kale, K.V.: Mapping several soil types using hyperspectral datasets and advanced machine learning methods. Results Opt. **12**, 100503 (2023). https://doi.org/10.1016/j.rio.2023.100503. ISSN 2666-9501

27. Hao, S., Wang, W., Salzmann, M.: Geometry-aware deep recurrent neural networks for hyperspectral image classification. IEEE Trans. Geosci. Remote Sens. **59**(3), 2448–2460 (2021). https://doi.org/10.1109/TGRS.2020.3005623

28. Barry, P.: EO-1/Hyperion Science Data User's Guide. TRW Space, Defense & Information Systems (2001)

29. Datt, B., McVicar, T.R., Van Niel, T.G., Jupp, D.L.B., Pearlman, J.S.: Preprocessing EO-1 hyperion hyperspectral data to support the application of agricultural indexes. IEEE Trans. Geosci. Remote Sens. **41**(6), 1246–1259 (2003). https://doi.org/10.1109/TGRS.2003.813206
30. https://www.nv5geospatialsoftware.com/Portals/0/pdfs/Confirmation/Hyperspectral-Whitepaper.pdf
31. https://pro.arcgis.com/en/pro-app/latest/tool-reference/image-analyst/train-deep-learning-model.htm

Novel Channel Fuzzy Logic System Modeling for Aquatic Acoustic Wireless Communication Within a Tank

Sweta Panchal[1] 📷, Sagar Patel[2](✉) 📷, and Sandip Panchal[1]

[1] Department of Electronics and Communications Engineering, SOET, Dr. Subhash University, Junagadh, Gujarat, India
[2] Department of Electronics and Communications Engineering, CSPIT, CHARUSAT, Changa, Gujarat, India
sagarpatel.phd@gmail.com

Abstract. This research explores underwater wireless communication in confined spaces, specifically water tanks, departing from the common focus on seawater in 5G networks. The study employs a unique system model using fuzzy logic techniques to analyze the impact of underwater applications. Unlike traditional sonar models, the approach emphasizes secured data transmission based on constraint statements, addressing signal attenuation in sodium chloride-saturated tanks. A tailored attenuation model and a multipath propagation model are developed to counter challenges like inter-symbol interference and protocol latency. Considering the tank's properties, a channel propagation model is designed for reliable communication. The study introduces a Wireless Channel propagation model to calculate signal coverage area. Using the Fuzzy Logic tool in MATLAB, simulations evaluate attenuation, multipath components, and coverage area for acoustic signals in the underwater frequency range. Fuzzy logic analysis provides comprehensive insights into underwater communication dynamics, specifically assessing the attenuation of acoustic frequency signals.

Keywords: 5G · wireless channel propagation model · underwater acoustic signal · acoustic sensor · fuzzy logic

1 Introduction

In today's technology-driven era, 5G wireless communication stands as the cornerstone. Its applications span across various sectors, including cellular communication, the Internet of Things (IoT), and wireless sensor networks [1–4]. Wireless sensor networks find applications in military, health, environmental monitoring, commercial fields and residential areas. While extensive research has been conducted on wireless communication and channel propagation models on land, underwater acoustic sensor networks (UAWSN) are gaining prominence for exploring aquatic environments, addressing issues like security and monitoring [5]. UAWSN comprises wireless sensors designed for collaborative tasks within specific areas, making it a valuable tool

Supported by CHARUSAT and Dr. Subhash University

for underwater communication challenges [6]. Unlike radio waves, which suffer from low signal-to-noise ratio, and high attenuation, and optical waves, which face scattering issues and limited range in water, acoustic waves prove to be the optimal choice for underwater communication. Acoustic sensors operate within a specific frequency range, making them effective for short-distance communication in aquatic environments, ranging from specific hertz to megahertz frequencies UAWSN, operating within this acoustic frequency range, offers promising solutions for communication hurdles, especially in confined spaces like tanks, where cleanliness and hygiene pose significant challenges [7–9].

In underwater tanks, the raw water composition includes salts like sodium chloride, boric acid, magnesium chloride and sulfuric acid, impacting acoustic signal absorption. Challenges in underwater audio communication arise from signal absorption, inhomogeneity, Ricean fading, and time-varying traits. Multipath propagation, caused by spreading, sediments, and chemical makeup, hinders signals. A 5G wireless communication channel model for enclosed underwater spaces is under development, addressing the shortcomings of existing underwater acoustic models. The study aims to fill the gap by creating a model tailored for 5G applications in underwater tanks, ensuring reliable communication [10, 25].

The paper is organized as follows. The Signal Attenuation and Transmission Method is presented in Sect. 2. The Propagation with Multiple Paths are shown in Sect. 3. Section 4 focus on creating an application channel model for Tank Deployment. Section 5 presents Simulation Results and discussions. Section 5 deals with the conclusions.

2   Signal Attenuation and Transmission Method

This section discusses the importance of investigating the properties of the underwater tank medium to develop an effective underwater acoustic wireless network. The goal is to create a propagation model that can accurately assess signal strength for reliable communication and facilitate the development of an automatic wireless communication system. Unlike deep-sea environments, tanks have distinct characteristics and chemical constituents, requiring a specific attenuation model. A sodium chloride-infused water attenuation model, common in tanks, has been formulated. The channel propagation model incorporates this attenuation model to calculate received power in underwater tanks. Due to multipath propagation, the model also estimates the number of multipaths based on communication distance. The proposed method, using fuzzy logic, aims to establish a communication system in tanks, catering to various shallow water applications.

2.1 Signal Attenuation and Transmission

The transmission of acoustic energy through a medium results in the attenuation of energy and a decrease in intensity, ultimately facilitating the propagation of acoustic waves. Path loss and spreading loss occur due to the sound signal from the transmitter spreading out in a geometric manner and diminishing in strength. Eq. (1) [11] gives the spreading loss from the source.

$$P_{spread} = K * 10log(R) \quad dB \tag{1}$$

The spreading factor, represented as K, is equivalent to the range in meters, represented as R. For unbounded transmission, K has a value of 2, whereas for bounded transmission, it has a value of 1 [11]. Absorption loss, indicated by Eq. (2), refers to the conversion of energy into heat because of Ionic relaxation and viscous friction during the propagation of the acoustic waves [11].

$$P_{absorb} = 10log(\alpha_{overall}) * R \quad dB \tag{2}$$

When denoting the range as R in meters and $\alpha_{overall}$ as the attenuation coefficient in the proposed model, the total path loss P_{total} Eq. (3) [11] combines the effects of spreading and attenuation losses.

$$P_{total} = P_{spread} + P_{absorb} dB \tag{3}$$

Equation (4) [11] determines the path losses' magnitude,

$$L = 10^{-(P_{total}/10)} \quad dB \tag{4}$$

The loss of signal strength, known as path loss, is predominantly influenced by factors like the depth of the medium used for communication, temperature, hydro-static pressure, and frequency. When the depth is below 100 m and the temperature fluctuates between 0°C and 20°C, the hydro-static pressure stays stable. Within this temperature span, attenuation increases in direct proportion to the square of the frequency. Previous studies have noted the occurrence of absorption losses in both shallow and deep water conditions [11].

2.2 Elements Found in Water

Ionic minerals such as calcium, sulfate, magnesium, iron, chloride, and bicarbonate can be found in raw water. A comparative examination of these ion proportions in sea and raw water is outlined in Table 1.

Table 1. Ion proportions in sea and raw water

Part of the composition	Ratio in parts per million (Sea water)	Ratio in parts per million (Raw water)
Chlorine	19346	2.376
Sodium	11,143	9.126
sulfate	2702	5.26
magnesium	1296	1.6
calcium	417	8.3
bromine	67	–
Borate	28	35.6
strontium cation	14	5.26
TOTAL(parts per million)	35151	63.26
TOTAL(percent)	3.516	0.006326
Pure water((percent)	96.486	99.993

The primary contributors to signal absorption in both sea water and raw water are magnesium sulfate, boric acid, and sodium chloride. In the provided sample, nearly 99.93% of the raw water consists of pure water. Among the constituents, sodium chloride accounts for 18.181% of the total ratio in the raw water. This specific chemical compound possesses unique properties that make it particularly effective in absorbing acoustic signals. Francosis and Garrison demonstrated the acoustic signal absorption in seawater by magnesium sulfate and boric acid [12, 13]. In the same way, every substance has a limited capacity to absorb sound waves in water.

2.3 Pure Water Attenuation Model

Equation (5) indicates that the signal absorption in pure water is caused by the material's sensitivity to pressure and particular relaxation frequencies.

$$P = 1 - 3.83 * 10^{-4}p + 4.9 * 10^{-8}p^2 \tag{5}$$

In the study conducted by Litovitz and Carnevale, the influence of pressure on signal absorption in pure water was investigated at temperatures of 0°C and 30°C [12]. A second-degree equation was fitted to data collected at 0°C due to the drop in temperature [12]. The researchers utilized an approximation equivalent to 1 atmosphere, corresponding to a water depth of 10 m. Here, 'z' denotes the depth in meters, while 'p' denotes the pressure in bars (p=z/10). The absorption equation was formulated accordingly.

$$P = 1 - 3.83 * 10^{-5}z + 4.9 * 10^{-10}z^2 \tag{6}$$

P is a bar representation of the hydrostatic pressure in Eq. (6). The water medium's temperature drops as depth increases. The temperature dependency was established using experimental data and, as described in the paper [14], was further separated into two segments for temperatures below and above 20°C. The temperature constants derived from this analysis are provided in Eq. (7) and Eq. (8).

For $T \leq 20°C$

$$A = 4.937 * 10^{-4} - 2.59 * 10^{-5}T + 9.11 * 10^{-7}T^2 - 1.50 * 10^{-8}T^3 \quad dBkm^{-1}khz^{-2} \quad (7)$$

For $T > 20°C$

$$A = 3.964 * 10^{-4} - 1.146 * 10^{-5}T + 1.45 * 10^{-7}T^2 - 6.5 * 10^{-10}T^3 \quad dBkm^{-1}khz^{-2} \quad (8)$$

The experimental data showed a significant increase in absorption magnitude with varying frequencies. Hence, the relationship between the attenuation coefficient, pressure, depth, frequency and temperature is described by Eq. (9) [14]. The hydro-static pressure "P" is linked to the medium's depth, and the temperature dependence coefficient is associated with the medium's temperature.

$$\alpha_{water} = P * A * f^2 * 10^{-3} \quad dB/m \quad (9)$$

The validity of the model has been established for a transmission medium with a maximum water temperature of 20°Celsius and a depth of 100 m., encompassing underwater signal frequencies from 10 Hz to 1 MHz.

2.4 Sodium Chloride Attenuation Model

The thermal conductivity of sodium chloride is constant at ambient temperature up to 25°C, at 0.5773 W/m°C [15]. Equation (10) expresses the relationship between attenuation, concentration of sodium chloride, heat conduction, and viscous losses.

$$\alpha = (\frac{\omega^2}{2\rho v^3})[(\frac{4\eta}{3}) + \eta^v + \tau(\frac{1}{c_p} - \frac{1}{c_v})] \quad dB/\mu s^{-1} \quad (10)$$

where α symbolizes attenuation in dB/m, f denotes frequency in Hz, ρ represents density in g/m^3, v denotes velocity in η^v, η represents volume viscosity, and η represents thermal conductivity in the given equation. The material's heat capacity at constant pressure and temperature is represented by the specific heat constants, c_p and c_v, respectively [16]. These parameters' values are calculated and used to determine the attenuation caused by sodium chloride [13]. The shear viscosity for sodium chloride concentrations between 2 and 3 mol/liter at 20°C and 1 bar of pressure is $\eta = 1048 \; \mu$ Pas, whereas the thermal conductivity is $\eta = 0.5773$ [16,17]. Equation (11) is given to represent the attenuation caused by sodium chloride.

$$\alpha_{Nacl} = 0.28f^2 \quad dB/m \quad (11)$$

According to Eq. (11), attenuation brought on by sodium chloride at a concentration of 3 mol/L shows frequency dependency. Attenuation increases with increasing frequency in shallow underwater tanks (maximum depth of 100 m), As a result, the attenuation coefficient and frequency squared are directly related.

2.5   A Model of Proposed Attenuation for Sodium Chloride in Water and Its Outcomes

An attenuation model has been created specifically for water saturated with sodium chloride, which is commonly used as the medium for transmitting underwater acoustic signals. The raw water, typically sourced from natural reservoirs, is composed mainly of 99.993% pure water, with a small fraction (0.00635%) consisting of other constituents, primarily sodium chloride. In natural aquatic environments, sodium chloride makes up 4.66% of the total concentration, with the remaining 95.34% being pure water. The model is expressed as a frequency-dependent equation in Eq. (12), as referenced in the literature [18].

$$\alpha_{overall} = 0.9534\alpha_{water} + 0.0466\alpha_{Nacl}   dB/meter \qquad (12)$$

$\alpha_{overall}$ denotes the attenuation coefficient for underwater sound waves in tanks that use raw water that has been saturated with sodium chloride as the medium for transmission. MATLAB has been used to develop and simulate the model for different acoustic frequency ranges at $0°C$ and $20°C$ temperatures.

This section discusses the simulation results of a proposed underwater acoustic communication model. Figure 1 displays attenuation levels at different frequencies in the range of 10 Hz to 1 MHz. The model shows lower attenuation rates compared to ultrasonic frequencies and is suitable for communication scenarios in water tanks with a maximum depth of 100 m, a hydrostatic pressure of 1 bar (dependent on communication frequency), and a temperature of $20°C$. The simulation considers temperatures of both $10°C$ and $20°C$. Increasing sodium chloride concentration leads to higher signal transmission losses, with absorption losses being negligible at around 0.5 mol/L but escalating at higher concentrations, resulting in decreased received signal strength due to increased attenuation loss and transmission loss.

2.6   A New Attenuation Model for Frequency and Variability in Range

The distance between the communication devices affects the acoustic signal's absorption loss. Equation (12) describes the suggested model, which is frequency-dependent and expressed in dB/meter. To explore changes in the attenuation coefficient and, consequently, absorption loss, the model has been extended to be range-reliant, as shown in Eq. (13).

$$\alpha_{overall} = (0.9534\alpha_{water} + 0.0466\alpha_{Nacl}) * L \qquad dB \qquad (13)$$

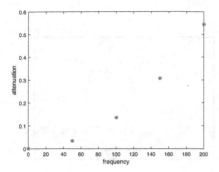

Fig. 1. Attenuation model simulated for various frequency at temperature 10°C and 20°C

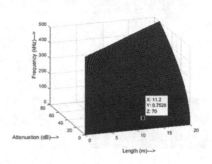

Fig. 2. Dependent on propagation range and frequency, the attenuation model

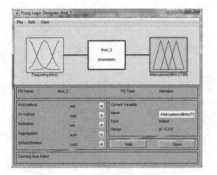

Fig. 3. Fuzzy logic is used in the FIS editor's input and output membership functions

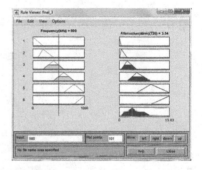

Fig. 4. Rule Viewer's fuzzy attenuation model

In this equation, $\alpha_{overall}$ represents the attenuation coefficient in decibels, α_{Nacl} and α_{water} denote the attenuation coefficients for underwater acoustic signals in sodium chloride concentrated raw water and, pure water respectively, and L represents the communication distance. The extended model in Eq. (13) has been simulated using the Fuzzy Logic tool in MATLAB, considering various acoustic frequencies and communication distances up to 20 m.

The simulated results of the frequency and range-dependent attenuation coefficient model are displayed in Fig. 2.

2.7 Applying Fuzzy Logic Techniques to Realize the Proposed Model

The MATLAB platform's Fuzzy Logic Toolbox, which consists of a number of functions in the MATLAB computer environment, is used to perform fuzzy logic operations. The membership rule viewer, FIS editor, rule base, surface viewer, and function editor are important tools for creating and adjusting fuzzy inference systems. Employing the fuzzy logic toolbox, the tuned fuzzy model aligned with the envisioned attenuation model in Eq. (12) is simulated. Within the fuzzy system's FIS editor, input parameters consist of frequency (in kHz), while the output variable is Attenuation (in dB/m). The fuzzification process is conducted utilizing the Mamdani inference engine.

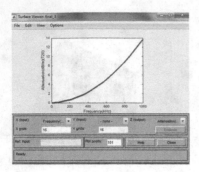

Fig. 5. Soft attenuation model (Surface observer)

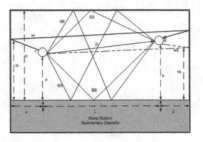

Fig. 6. Fuzzy logic in the water route and tank shape causes multipath signals to propagate

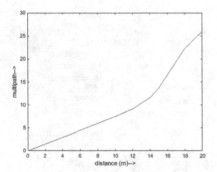

Fig. 7. surface viewer's fuzzy attenuation model

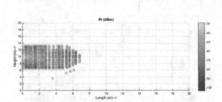

Fig. 8. Water channel and tank geometry with fuzzy logic causing multipath signal propagation

In Fig. 3, the fuzzy inference engine corresponding to the envisaged attenuation model in Eq. (12) is depicted. In this case, frequency and attenuation serve as input and output variables, respectively, and are considered input parameters. The Mamdani inference technique is employed by the Fuzzy Inference System (FIS), while the mean of maximum approach is applied by the defuzzification method.

2.8 Viewer for Rules and Viewer for Surfaces

Figure 5 shows the surface viewer of the fuzzy attenuation model, where attenuation is the output variable and frequency is the input variable. Figure 4 shows the rule viewer.

The proposed model is approximately simulated utilizing soft computing techniques such as Fuzzy Logic.

3 Propagation with Multiple Paths

Many reflections of the transmitted acoustic signal are produced by the physical characteristics and structure of the communication environment. Multi-path signals are created as a result of these reflections and travel in the direction of the receiver. The transmitted acoustic signal consists of many route segments in addition to the principal signal. Every incoming signal, arriving from different pathways with different propagation delays, is detected by the receiver.

3.1 Geometry of the Underwater Acoustic Channel in Tanks

The primary multi-path signals combine constructively to enhance signal strength at the receiver [19].

The multipath signals produced by the acoustic sound reflecting off the tank walls are shown in Fig. 6. After being sent, the signal travels towards the receiver after striking the wall behind it at a certain height h_i, reflecting off of it. The possible reflection from the wall behind the receiver is shown in Fig. 4. As an alternative, the signal may travel away from the receiver and reflect off the tank wall at a height of h_k. In this case, W_i denotes the signal that is reflected off the wall behind the transmitter, and W_k denotes the signal that is traveling in front of the receiver.

3.2 Path Lengths

Multiple reflections of the transmitted acoustic signal occur as a result of the physical qualities and characteristics of the communication medium. The multipath signals that head in the direction of the receiver are largely produced by these reflections. The direct signal and multipath signal path lengths in the tank, as shown in Fig. 6, are given by the following Eqs. (14) to (15) [20], where 'n' denotes the order of reflection.

$$D = \sqrt{L^2 + (b-a)^2}, SS = \sqrt{L^2 + (2nH - a - b)^2}, SB = \sqrt{L^2 + (2nH - a + b)^2} \quad (14)$$

$$BS = \sqrt{L^2 + (2nH + a - b)^2}, BB = \sqrt{L^2 + (2(n-1)H + a + b)^2} \quad (15)$$

Equations (16) to (19), extracted from Fig. 6, provide the path lengths of signals reflected from the tank walls.

$$W_k = \sqrt{(L+\beta)^2 + (a - nh_k)^2} + \sqrt{(b - nh_k)^2 + \beta^2} \quad if \quad h_k < a < b \quad (16)$$

$$W_k = \sqrt{(L+\beta)^2 + (a + nh_k)^2} + \sqrt{(b - nh_k)^2 + \beta^2} \quad if \quad a < h_k < b \quad (17)$$

$$W_i = \sqrt{\alpha^2 + (a - nh_i)^2} + \sqrt{(\alpha + 1)^2 + (b - nh_i)^2} \quad if \quad h_i < a \quad (18)$$

$$W_i = \sqrt{\alpha^2 + (a + nh_i)^2} + \sqrt{(\alpha + 1)^2 + (b - nh_i)^2} \quad if \quad h_i > a \quad (19)$$

By employing Eqs. (14) to (19), one can compute the path lengths for both the direct signal and multipath signals. Utilizing these path lengths allows the calculation of propagation delays.

3.3 Delay in the Propagation of Multipath Signals

The propagation delay for the reflected signal is the amount of time that passes between the arrival time of the transmitted signal and its reflection time. As described in the reference [20], this delay can be calculated using Eqs. (20), which take surface-to-bottom, bottom-to-surface, and surface and bottom reflections into consideration.

$$\tau_{SB} = \frac{SB - D}{C}, \tau_{BS} = \frac{BS - D}{C}, \tau_{SS} = \frac{SS - D}{C}, \tau_{BB} = \frac{BB - D}{C}, \tau_{W_k} = \frac{W_k - D}{C}, \tau_{W_i} = \frac{W_i - D}{C} \quad (20)$$

The propagation delay plays a crucial role in underwater acoustic signal transmission, as it causes multiple receptions of information at the receiver due to the delayed arrival of multi-path signals. To assess the stability of propagation delays, differentiation of these delays concerning the distance (L), receiver depth (b), and transmitter depth (a), between communicating devices (transmitter and receiver) can be conducted.

3.4 Multipath Components in Communication Range

The mean-square error as a function of the maximum frequency interval, determined using five approaches, is shown in Eq. (21) [21].

$$V_{max} = \frac{\delta - 1}{2\tau_{max}} \quad \bullet \quad (21)$$

In practical scenarios, the number of discrete paths, denoted by δ and the maximum propagation delay τ_{max} are essential parameters. The total bandwidth B of the transmitted acoustic signal is represented by V_{max}. Consequently, the number of multi-paths is δ can be calculated using Eq. (22) [21].

$$\delta = [2\tau_{max}B] + 1 \quad (22)$$

The significance of the maximum propagation delay (τ_{max}) of multi-path components is given in the evaluation of acoustic signals. The evaluation involves considering bandwidth (B in kilohertz) and analyzing transmission ranges of 10 m and 20 m. Arrival time discrepancies are calculated based on path lengths in both dimensions. Multi-path components are identified through the maximum propagation delay, showing an increase in multi-paths as transmission ranges extend. The pattern of exponential growth in the number of multi-paths is observed up to 20 m, as indicated by the relationship expressed in Eq. (23) [22].

$$\delta = 4.17e_{0.224}L \quad (23)$$

Figure 7 illustrates the exponential growth in the number of multi-path components as the transmission range expands, primarily due to mobile communicating devices. This escalation in the communication range leads to increased reflections and consequently, more multi-paths. Hence, the increase in multiple signal paths restricts the maximum permissible distance between the transmitter and receiver to ensure precise signal reception within an acceptable propagation delay range. The research by [23] scrutinized the statistical properties of shallow water channels in diverse settings, including water tanks, swimming pools and lakes. The outcomes confirmed that a time-varying channel model based on Weibull or Ricean distribution fading patterns was appropriate.

4   Creating an Application Channel Model for Tank Deployment

Acoustic communication, like radio waves, follows the power intensity formula $P = \frac{P_t}{4\pi L^2}$, where P is received power intensity. The frequency range of omni-directional radiators in acoustics is lower than that of radio signals. Considerations for factors like transmitted power P_t, receiver distance L, and line-of-sight propagation are important, with attention to other radiation angles. Models such as Rayleigh or Ricean channels, similar to radio communication, are applied in acoustic systems for translating radio channel principles. The Friis transmission equation is used for estimating transmission power. [24, 25]

Radiated energy in underwater acoustic wireless sensor networks is omnidirectional. The received power (P) is a crucial metric, expressed by Eq. (24). Here, P_r is the transmitted power, and G_t and G_r are gains for the transmitter and receiver. λ is the wavelength at the operational frequency, and L is the communication distance.

$$P = \frac{P_r G_t G_r \lambda^2}{16\pi^2 L^2} \tag{24}$$

The received power, accounting for reflection coefficient (Γ) and phase shift (ϕ) from the reflected signal, is determined by Eq. (25). The acoustic signal travels at 1500 m/s in water, and the radiating device uses horizontal polarization.

$$P = \frac{P_t G_t G_r \lambda^2}{16\pi^2 L^2} |1 + \Gamma \exp j\phi|^2 \tag{25}$$

Equation (26) [26] is used to calculate the Reflection coefficient, which is generated by the impedance mismatch between water and air. The analogous f_2 at 9.7 knots wind speed is 4.017 kHz, and Γ is fixed at 0.49.

$$|r_s| = \sqrt{\frac{(1 + (f/f_1)^2)}{(1 + (f/f_2)^2)}}, where f_2 = 378w^{-2}, f_1 = \sqrt{10}f^2 \tag{26}$$

Equation (27) gives the reflection coefficient of the angle-dependent reflected signal in a tank with a smooth bottom. The NSUC model is used to calculate the reflection coefficient for the bottom-reflected wave [27].

$$|r_b| = \frac{m\cos\theta - \sqrt{n^2 - \sin\theta}}{m\cos\theta + \sqrt{n^2 - \sin\theta}} \tag{27}$$

Diffraction loss arises when an obstacle taller than the wavelength obstructs the direct communication wave path. Signal strength is greater when the signal bends behind a sharp-edged obstacle compared to a rounded one [28]. However, in the chosen communication medium, a tank with a smooth bottom, such diffraction effects are not observed. Scattering losses due to reflection are minimal in this medium, as confirmed by Gibson's observations [29].

4.1 Underwater Tank Propagation Model for Acoustic Signals

The given section outlines the development of a comprehensive model for underwater acoustic communication, specifically tailored for underwater tanks. The model takes into account various propagation losses such as path loss, reflection losses, and attenuation loss. It is designed to operate in a medium with constant temperature and pressure, where reflections occur due to the tank's geometry and water channel arrangement, resulting in multi-path propagation. The section introduces an equation (Eq. (23)) to quantify the number of multi-paths in relation to the communication range and provides a mathematical representation of the underwater tank channel propagation model in Eq. (28).

$$P_r = \frac{P_t G_t G_r \lambda^2}{16\pi^2} |\alpha_{overall} \frac{1}{d_d} \exp(-jkd_d) + \Gamma_1(f) \frac{1}{d_1} \exp(-jkd_1)...$$
$$... + \Gamma_2(f) \frac{1}{d_2} \exp(-jkd_2) + \Gamma_n(f) \frac{1}{d_n} \exp(-jkd_n)|^2 \tag{28}$$

The suggested attenuation for a water tank saturated with sodium chloride, as determined by Eq. (12), is denoted by $\alpha_{overall}$ in this context. In this case, the direct communication path is denoted by d_d, while the numerous paths (total of δ paths) are d_1, d_2,..., d_n. The most multi-paths that can exist within a given communication range are defined by Eq. (23). $d_1, d_2,..., d_n$ lengths, up to δ paths, correspond to multi-paths such as SS, BB, BS, SB, and so on. The reflection loss of the nth wave at the operational frequency $f = 30$ kHz is denoted by the expression $\Gamma_n(f)$, where k is the phase constant. The $exp(-jkd_n)$ component of each multi-path signal determines its phase. The distances from the transmitter and receiver to the reflected signal are added together to find d_n, which is the distance covered by the nth ray and the accompanying reflection coefficient.

The bounded area coverage for a 20×20 meter square site is 31.9%, emphasizing its importance. Placing acoustic sensor nodes strategically within a water tank is crucial for maximizing power reception, based on areas where receiving sensitivity exceeds -80 dBm. MATLAB's Fuzzy Logic tool simulates channel propagation in the undersea tank with a 30 kHz sensor that has omnidirectional radiation, a 10 kHz bandwidth, and 0 dBm power in order to find the best location for the sensor node for effective power reception.

Figure 8 depicts simulation results of a channel propagation model, illustrating received power at various locations in a tank. It aids in determining optimal sensor placement for coverage. Results show decreased transmitted power with increasing communication distance.

5 Result and Discussion

Figure 1 displays simulation results showcasing the proposed attenuation model's performance across different frequencies and temperatures. The study focuses on communication in a shallow water tank (depth < 100 m) with a constant hydrostatic pressure of 1 bar. The sodium chloride concentration is maintained at 0.5 mol/L for dependable communication with minimal transmission loss.

Higher concentrations of sodium chloride and increased temperature in the communication medium lead to a rise in signal attenuation. This effect is especially pronounced at higher ultrasonic frequencies. The proposed attenuation model is crucial for evaluating the signal strength at the receiver end.

The strength of a transmitted signal is influenced by factors such as sodium chloride concentration, communication distance, transmission frequency, and temperature due to their impact on multipath propagation. These factors act as constraints, affecting the signal's strength and coverage. The channel propagation model helps identify the signal's coverage area, facilitating accurate reception and ensuring faithful signal transmission in a tank based on medium characteristics and sodium chloride concentration.

The proposed attenuation model is suitable for underwater tanks with a sodium chloride concentration of 0.5 mol/L. However, in shallow tanks, there are additional chemicals, including sulphates, calcium, boric acid, etc. (Table 1), which significantly impact signal strength. Existing models are inadequate for detecting signal attenuation in the presence of these chemicals. A new model is needed to account for all water constituents. The proposed channel propagation model calculates received power for signals transmitted in sodium chloride-saturated water. A similar model is required for other chemicals like boric acid, sulphates, calcium, etc., in the underwater medium.

Conclusions

The MATLAB simulation using a Fuzzy logic tool indicates that increasing concentrations of sodium chloride and acoustic frequency lead to higher attenuation of transmitted acoustic signals. Specifically, there is significant attenuation in the ultrasonic frequency range. The Fuzzy logic simulation confirms that attenuation losses rise with increased acoustic frequencies. Additionally, a multipath signal model reveals the number of multipath signals as a function of communication distance in a bounded underwater medium. The channel propagation model predicts the power received by a receiver at different locations in an underwater tank for a steady transmitter. Simulation results of the wireless channel propagation model illustrate the coverage area of the transmitted acoustic signal. This information can aid in determining the location of a receiver for reliable communication in the bounded underwater medium of 5G wireless communication.

References

1. Ali, M.F., Jayakody, D.N.K., Chursin, Y.A., Affes, S., Dmitry, S.: Recent advances and future directions on underwater wireless communications. Arch. Comput. Methods Eng. 27, 1379–1412 (2020)
2. Patel, A., et al.: UWB CPW fed 4-port connected ground MIMO antenna for sub-millimeter-wave 5G applications. Alex. Eng. J. 61(9), 6645–6658 (2022)
3. Patel, S., Patel, R., Bhalani, J.: Performance Analysis & implementation of different modulation techniques in Almouti MIMO scheme with Rayleigh channel. In: International Conference on Recent Trends in Computing and Communication Engineering (2013)
4. Patel, S.B., Bhalani, J., Trivedi, Y.N.: Performance of full rate non-orthogonal STBC in spatially correlated MIMO systems. Radioelectron. Commun. Syst. 63, 88–95 (2020)

5. Saraswala, P.P., Patel, S.B., Bhalani, J.K.: Performance metric analysis of transmission range in the ZigBee network using various soft computing techniques and the hardware implementation of ZigBee network on ARM-based controller. Wireless Netw. **27**(3), 2251–2270 (2021)
6. Patel, S., Dwivedi, V.V., Kosta, Y.P.: A parametric characterization and comparative study of Okumura and Hata propagation-loss-prediction models for wireless environment. Int. J. Electr. Eng. Res. **2**(4), 453–463 (2010)
7. Lurton, X.: An introduction to underwater acoustics: principles and applications. Noise Control Eng. J. **59**(1), 106 (2011)
8. Jiang, Z.: Underwater acoustic networks-issues and solutions. Int. J. Intell. Control Syst. **13**(3), 152–161 (2008)
9. Stojanovic, M.: Underwater Acoustic Communication. Wiley Encyclopedia of Electrical and Electronics Engineering, John Wiley and Sons (2015)
10. Porter, M.B.: Beam tracing for two-and three-dimensional problems in ocean acoustics. J. Acoust. Soc. Am. **146**(3), 2016–2029 (2019)
11. Akyildiz, I.F., Pompili, D., Melodia, T.: Underwater acoustic sensor networks: research challenges. Ad Hoc Netw. **3**(3), 257–279 (2005)
12. Francois, R.E., Garrison, G.R.: Sound absorption based on ocean measurements, part II: Boric acid contribution and equation for total absorption of sound. J. Acoust. Soc. Am. **72**, 1879–1890 (1982)
13. Francois, R.E., Garrison, G.R.: Sound absorption based on ocean measurements: part I: pure water and magnesium sulfate contributions. J. Acoust. Soc. Am. **72**(3), 896–907 (1982)
14. Burrowes, G., Khan, J.Y.: Short-range underwater acoustic communication networks. Autonomous Underwater Veh., 173–198 (2011)
15. Philips, S.L., Igbene, A.: A Technical databook for geothermal energy utilization. Lawrence Berkeley Laboratory, vol. 10 (1981)
16. Aleksandrov, A.A., Dzhuraeva, E.V., Utenkov, V.F.: Viscosity of aqueous solutions of sodium chloride. High Temp. **50**(3), 354–358 (2012)
17. Ozbek, H.: Viscosity of aqueous sodium chloride solution from 0–150 C. Lawrence Berkeley Laboratory (2010)
18. Panchal, S.S., Pabari, J.P.: Evaluation of shallow underwater acoustical communication model for attenuation and propagation loss for aqueous solution of sodium chloride. In: 2019 International Conference on Recent Advances in Energy-efficient Computing and Communication (ICRAECC), pp. 1–5. IEEE (2019)
19. Vallis, G.K.: Atmospheric and Oceanic Fluid Dynamics, Fundamental and Large Scale Circulation. Cambridge University Press, Cambridge, vol. 842, no. 2 (1995)
20. Zielinski, A., Yoon, Y.H., Wu, L.: Performance analysis of digital acoustic communication in a shallow water channel. IEEE J. Oceanic Eng. **20**(4), 293–299 (1995)
21. Patzold, M., Szczepanski, A., Youssef, N.: Methods for modeling of specified and measured multipath power-delay profiles. IEEE Trans. Veh. Technol. **51**(5), 978–988 (2002)
22. Panchal, S.S., Dwivedi, V.V., Pabari, J.P.: Design of attenuation loss and multipath propagation model for underwater acoustic communication in tank. Int. J. Recent Technol. Eng. (IJRTE) (2020)
23. Kulhandjian, H., Melodia, T.: Modeling underwater acoustic channels in short-range shallow water environments. In: Proceedings of the 9th International Conference on Underwater Networks & Systems, pp. 1–5 (2014)
24. Paul Van Walree, P.A.: Propagation and scattering effects in underwater acoustic communication channels. IEEE J. Oceanic Eng. **38**(4), 614–631 (2013)
25. Huang, J.G., Wang, H., He, C.B., Zhang, Q.F., Jing, L.Y.: Underwater acoustic communication and the general performance evaluation criteria. Front. Inf. Technol. Electr. Eng. **19**(8), 951–971 (2018)

26. Coates, R.: An empirical formula for computing the Beckmann-Spizzichino surface reflection loss coefficient. IEEE Trans. Ultrason. Ferroelectr. Freq. Control **35**(4), 522–523 (1988)
27. Yoon, Y.H.: High-rate digital acoustic communications in a shallow water channel (Doctoral dissertation) (1999)
28. Wong, H.: Field strength prediction in irregular terrain-the PTP model. Report of Federal Communication Commission, USA (2002)
29. Gibson, J.D.: The Communications Handbook. CRC Press, Boca Raton (2018)

Iterative Interference Cancellation for STBC-OFDM System Over Doubly Selective Channel

Jyoti P. Patra[1]([✉]) [iD], Bibhuti Bhusan Pradhan[2] [iD], Ranjan Kumar Mahapatra[3] [iD], and Sankata Bhanjan Prusty[4] [iD]

[1] Vidya Jyothi Institute of Technology, Hyderabad, India
jyotiprasannapatra@gmail.com
[2] Malla Reddy Engineering College, Hyderabad, India
[3] Koneru Lakshmaiah Education Foundation, Vijayawada, India
[4] REVA University, Bangalore, India

Abstract. The space time block coded orthogonal frequency division multiplexing (STBC-OFDM) system performance is severely degraded due to occurrence of co-channel interference (CCI) and inter-carrier interference (ICI) effects in mobile environment. In this work, three joint CCI and ICI iterative interference cancellation techniques are proposed namely diagonalized zero forcing detection (DZFD)-parallel interference cancellation (PIC)-DZFD, order iterative decision feedback (OIDF)-PIC-DZFD and OIDF-PIC-OIDF to obtain the transmitted signal over doubly selective channel. These proposed methods cancel the interferences in three stages. The CCI cancellation is performed to obtain initial data symbol in the initial stage followed by parallel-interference-cancelation (PIC). In the third stage, the ICI free signal is processed again by CCI cancellation method to obtain the refined estimated data symbols. Finally, the proposed and conventional methods are compared with respect to complexity and symbol error rate (SER). From the results, it is demonstrated that the proposed OIDF-PIC-OIDF method significantly outperforms the conventional methods with lower complexity.

Keywords: STBC-OFDM · Signal detection · Inter-carrier interference (ICI) · Co-channel interference (CCI) · Doubly selective channel

1 Introductions

In modern days, the transmit diversity techniques have achieved significant interest because they provide reiability without additional increses in bandwidth and transmit power [1, 2]. Space time block code (STBC), often known as transmit diversity, was first proposed in [3]. The STBC schemes become the popular wireless communication techniques for a variety of applications including LTE, WiMAX, Wi-Fi [5] and design of radio receivers for military and civilian users [4]. The STBC technique is suitable for flat fading channels. In reality, the channel experiences time and frequency selective. To overcome the frequency selective issues, the STBC scheme can be combined with

K. K. Patel et al. (Eds.): icSoftComp 2023, CCIS 2031, pp. 122–136, 2024.
https://doi.org/10.1007/978-3-031-53728-8_10

OFDM modulation technique [6]. However, STBC-OFDM technique severely degrades in fast fading chennel due to the occurance of inter-carrier interference (ICI) and co-channel interferences (CCI) [7–14]. The CCI effect arises because of channel frequency response (CFR) variation for consecutive time periods. The orthogonality loss between subcarriers in OFDM modulation causes the ICI effects [15–17]. In literature, a number of signal detecting techniques have been suggested by suppressing CCI and ICI effects. In [7], successive interference cancellation (SIC) based signal detection technique was proposed. However, it has a problem with error propagation. In [8], list-SIC method was suggested which suppress the error floor in SIC and thus significantly boosts system performance. In [9], to cancel the CCI effects, diagonalized zero forcing detection (DZFD) was proposed. In [10], decision feedback (DF) technique was discussed. The maximum-likelihood (ML) technique was discussed in [11–13]. The ML technique has the highest computational cost but achieves the best performance. In [11–13], several low complexities and close to ML methods have been suggested. In [11], QR decomposition with ordering based signal detection was proposed. Spatial permutation modulation scheme based signal detection was proposed in [12]. A modified zero foring (M-ZF) signal detection method was proposed for STBC-OFDM system in accoustic channel [13]. An ordered iterative decision feedback (OIDF) signal detection method was proposed in [14]. All these signal detection techniques just suppress the CCI effects to obtain desired signals. To further enhance the performance of system, it requires supression of both CCI and ICI effects simultaneously. In literature, several methods have been addressed by combating both CCI and ICI effects in [18–25]. In [18], based on MMSE windowing technique time domain block linear filter (TDBLF) was proposed. However, this method is computationally intensive due to large matrix inversion. A low complexity frequency domain block linear filter (FDBLF) was proposed in [19] which has lower computational complexity as compared to TDBLF. However, the performance of FDBLF is poorer than TDBLF. In [20], sequential decision feedback sequence equalization (SDFSE) technique was proposed. In [21], SAGE based signal detection method was proposed for STBC-OFDM system in mobile environment. An ordered block decision feedback equalizer (OBDFE) was proposed in [22]. These signal detection methods proposed in [18–22] have high complexity as they involve in matrix inversion. An iterative interference cancellation method was proposed in [23]. In [24], an interference suppression approach was addressed which was based on optimal selection of data symbol pair ordering. To cancel both CCI and ICI effects concurrently, a low overhead interference cancellation method was developed in [25], however it is inferior to ML method. Recently, several joint decoding and channel estimation techniques are proposed for STBC-OFDM in [26, 27]. A joint decoding and channel estimation technique based on complex-valued neural networks (CVNNs) was proposed in [26]. An excepatation maximization based joint channel estimation and detection was proposed for STBC-OFDM system in [27]. This method provides near to ML method with few pilot subcarriers.

In this paper, three low complexity signal detection techniques namely DZFD-PIC-DZFD, OIDF-PIC-DZFD and OIDF-PIC-OIDF are proposed for STBC OFDM scheme. The performances of proposed and conventional methods are compared with respect to complexity and symbol error rate (SER).

The remaining of the paper is organized as follows. In Sect. 2, we discuss the STBC-OFDM system followed by OIDF signal detection method. The various proposed joint iterative cancellation methods are presented in Sect. 3. We calculate the complexity of proposed and conventional methods in Sect. 4 and present the symbol error rate performances comparison in Sect. 5. In Sect. 6, we conclude the paper.

2 System Model

This scetion describes the STBC-OFDM system model follwed by the OIDF signal detection technique.

2.1 STBC-OFDM System Model

We consider an STBC-OFDM system for two transmitting antennas and single receiving antenna. At transmitting end, binary data sequence is generated, mapped and forwarded to the STBC block. The STBC transforms the mapped data into coded data symbol matrix $X(k)$ as per Alamouti encoding [3] and is given as follow

$$X(k) = \begin{bmatrix} X_{(1)}^1(k) \, X_{(1)}^2(k) \\ X_{(2)}^1(k) \, X_{(2)}^2(k) \end{bmatrix} = \begin{bmatrix} X_1(k) & X_2(k) \\ -X_2^*(k) & X_1^*(k) \end{bmatrix} \tag{1}$$

where $X_{(t)}^i(k)$ is k^{th} subcarrier for i^{th} transmitting antenna at t^{th} time before the IFFT operation. The transmitted signal after IFFT operation is given by

$$x_{(t)}^i(n) = \frac{1}{\sqrt{N}} \sum_{k=0}^{N-1} X_{(t)}^i(k) e^{\frac{j2\pi kn}{N}} \tag{2}$$

The time domain signal are then transmitted after adding cyclic prefix (CP). At receiver, the received signal is obtained as convolution operation bewteen channel impulse response (CIR) and transmitted signal as given below

$$y_{(t)}(n) = \sum_{i=1}^{2} \sum_{l=0}^{L-1} h_{(t)}^i(n, l) x_{(t)}^i(n - l) + w_{(t)}(n) \tag{3}$$

where $h_{(t)}^i(n, l)$ is the CIR for l-th channel tap during n-th sampling instant. The symbol $w_t(n)$ is additive white Gaussian noise (AWGN). The FFT operation is performed after the CP has been rempved and is written as

$$Y_{(t)}(k) = \sum_{i=1}^{2} \left[H_{(t)}^i(k, k) X_{(t)}^i(k) + I_{(t)}^i(k) \right] + W_{(t)}(k) \tag{4}$$

$$I_{(t)}^i(m) = \sum_{\substack{m=0 \\ m \neq k}}^{N-1} H_{(t)}^i(k, m) X_{(t)}^i(m) \tag{5}$$

$$H_{(t)}^i(k,m) = \frac{1}{N}\sum_{m=0}^{N-1}\sum_{l=0}^{L-1} h_{(t)}^i(n,l)e^{\frac{-j2\pi n(k-m)}{N}} e^{\frac{-j2\pi ml}{N}} \tag{6}$$

The received signal can be elaborated in matrix form as

$$Y(k) = \begin{bmatrix} Y_{(1)}(k) \\ -Y_{(2)}^*(k) \end{bmatrix} = \underbrace{\begin{bmatrix} H_{(1)}^1(k,k) & H_{(1)}^2(k,k) \\ H_{(2)}^{2*}(k,k) & -H_{(2)}^{1*}(k,k) \end{bmatrix}}_{H(k,k)} \underbrace{\begin{bmatrix} X_1(k) \\ X_2(k) \end{bmatrix}}_{X(k)}$$

$$+ \underbrace{\sum_{m=0m\neq k}^{N-1} \begin{bmatrix} H_{(1)}^1(k,m) & H_{(1)}^2(k,m) \\ H_{(2)}^{2*}(k,m) & -H_{(2)}^{1*}(k,m) \end{bmatrix}\begin{bmatrix} X_1(m) \\ X_2(m) \end{bmatrix}}_{I(k)} + \underbrace{\begin{bmatrix} W_{(1)}(k) \\ W_{(2)}^*(k) \end{bmatrix}}_{W(k)}$$

$$\tag{7}$$

The symbols $Y(k)$, $W(k)$ and $I(k)$ are received signal, AWGN, and interferences signal respectively. The effect of ICI is initially ignored and it can be considered as Gaussian process [17]. Hence, the received signal is modified as

$$Y(k) = H(k,k)X(k) + J(k) \tag{8}$$

$$J(k) = I(k) + W(k) \tag{9}$$

After performing multiplication operation between $H^H(k,k)$ with $Y(k)$, the estimated signal is written as

$$\tilde{X}(k) = H^H(k,k)Y(k) = G(k)X(k) + H^H(k,k)J(k) \tag{10}$$

$$G(k) = H^H(k,k) \times H(k,k) = \begin{bmatrix} \alpha_1(k) & \beta(k) \\ \beta^*(k) & \alpha_2(k) \end{bmatrix} \tag{11}$$

$$\alpha_1(k) = \left|H_{(1)}^1(k)\right|^2 + \left|H_{(2)}^2(k)\right|^2, \alpha_2(k) = \left|H_{(2)}^1(k)\right|^2 + \left|H_{(1)}^2(k)\right|^2 \tag{12}$$

$$\beta(k) = H_{(1)}^1 * (k) H_{(1)}^2(k) - H_{(2)}^1 * (k) H_{(2)}^2(k) \tag{13}$$

where $\alpha_1(k)$, $\alpha_2(k)$ are diversity gain term. The symbols $\beta(k)$, $\beta^*(k)$ are the unwanted CCI terms. The estimated signal $\tilde{X}(k)$ is exprseed in matrix form

$$\tilde{X}(k) = \begin{bmatrix} \tilde{X}_1(k) \\ \tilde{X}_2(k) \end{bmatrix} = \begin{bmatrix} \alpha_1(k)X_1(k) + \beta(k)X_2(k) + Z_1(k) \\ \beta^*(k)X_1(k) + \alpha_2(k)X_2(k) + Z_2(k) \end{bmatrix} \tag{14}$$

where $\alpha_1(k)X_1(k)$ and $\alpha_2(k)X_2(k)$ are diversity signals. $\beta(k)X_2(k)$ and $\beta^*(k)X_1(k)$ denote unwanted interferences signals. These CCI signals are mixed with original signals, thus degrades system performance. To improve performance of system, several methods have been proposed such as SIC, DZFD, DF, ML and OIDF. The ML signal detection technique provides optimal performance with high complexity. However, the order iterative decision feedback (OIDF) technique provides similar performance as ML with significaly reduced complexity.

2.2 Order Iterative Decision Feedback (OIDF) Method

The OIDF technique was proposed in [14]. The OIDF technique first obtains initial estimated signal using DZFD method and calculates the unwanted interreference signal. Later, it iteratively cancels the interreference to obtain refined data signal. The algorithm for OIDF method is illustrated below

Initialization: Apply DZFD Method

Since, multiplying $H^H(k, k)$ with the received signal Y(k) is not a diagonal matrix, $\Omega(k, k)$ matrix is multiplied in place of $H^H(k, k)$ and can be expressed as

$$\tilde{X}(k) = \Omega Y = diag(\phi, \phi)X + \Omega J \tag{15}$$

The transform matrix $\Omega(k)$ after simplification of (15) is given by

$$\Omega(k, k) = \begin{bmatrix} H_{21}^*(k, k) & H_{12}(k, k) \\ H_{22}^*(k, k) & -H_{11}(k, k) \end{bmatrix} \tag{16}$$

The value of $\phi(k)$ is obtained after multiplication of Ω and H matrix

$$\phi(k) = H_{11}(k)H_{21}^*(k) + H_{12}(k)H_{22}^*(k) \tag{17}$$

The estimated signal is obtained by dividing $\phi(k)$ with $\tilde{X}(k)$ as given in (15) followed by hard decision operation

$$\hat{X}(k) = Q\left(\frac{\tilde{X}(k)}{\phi(k)}\right) = X(k) + J(k) \tag{18}$$

where Q denotes hard decision function.

OIDF Algorithm:

[*Step* 1]: Set the variables as based on diversity gain

If $\alpha_1(k) \geq \alpha_2(k)$, set variable $a = 1$ and $b = 2$

If $\alpha_1(k) < \alpha_2(k)$, set variable $a = 2$ and $b = 1$

[*Step* 2]: Detect the data signal using DZFD output with larger diversity gain

$$\hat{X}_a(k) = Q((\tilde{X}_a(k)/\phi(k)) = X_a(k) + J(k) \tag{19}$$

[*Step* 3]: Detects the data signal through iteratively cancel the CCI effects

$$\text{for } I = 1 : P \tag{20}$$
$$\hat{X}_b(k) = Q\{(\tilde{X}_b(k) - \beta^*\hat{X}_a(k)/\rho_b(k)\}$$

$$\hat{X}_a(k) = Q\{(\tilde{X}_a(k) - \beta\hat{X}_b(k)/\rho_a(k)\} \tag{21}$$
$$\text{end}$$

where P denotes number of iterations.

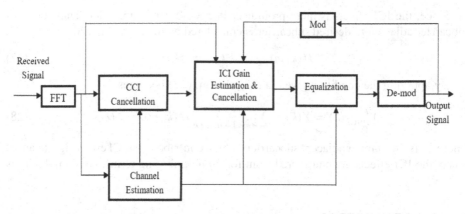

Fig. 1. Block diagram of proposed joint CCI-ICI cancellation method

3 Iterative Interference Cancellation Methods

This section describe the proposed iterative cancellation methods namely DZFD-PIC-DZFD, OIDF-PIC-DZFD and OIDF-PIC-OIDF are presented. Figure 1 illustrates working principles of propsed signal detection methods.

3.1 DZFD-PIC-DZFD Method

The DZFD-PIC-DZFD method cancels the interferences iteratively as shown in Fig. 1. This method estimates the transmitted signal using three stages. In first stage, DZFD technique is adopted to get initial signal. In second stage, ICI signals are calculated and removed from received signal as given below

$$Y_{offICI}^I(k) = Y(k) - \sum_{m=0,m\neq k}^{N-1} H(k,m)\hat{X}^{I-1}(m) = H(k,k)\hat{X}^I(k) + W(k) \quad (22)$$

$$Y_{offICI}^I(k) = \left[Y_{(1)offICI}^I(k) \left(Y_{(2)offICI}^I(k) \right)^* \right] \quad (23)$$

where $\hat{X}^{I-1}(m)$ is the estimated signal obtained from DZFD in (18). In third stage, the DZFD method is again applied to ICI free signal as follows

$$R_{DZFD}^I(k) = \Omega(k,k) \times Y_{offICI}^I(k) \quad (24)$$

$$R_{DZFD}^I(k) = \left[R_{(1)DZFD}^I(k) \ R_{(2)DZFD}^I(k) \right] \quad (25)$$

The estimated signal is obtained after performing the hard decision function of (25) and is expressed as

$$\hat{X}^I(k) = Q\left(R_{DZFD}^I(k) \right) = X(k) + W'(k) \quad (26)$$

Since, the ICI effects is more prominent in neighboring subcarrier. Thus, only $2q$ subcarrier adjacent to desired subcarrier is considered as mentioned in [20]

$$H(k, m) = 0 \, for \, |k - m| > q \tag{27}$$

By applying the condition given in (27), the Eq. (22) becomes

$$Y_{offICI}^{I}(k) = Y(k) - \sum_{m=k-q, m \neq k}^{m=k+q} H(k, m)\hat{X}^{I-1}(m) \tag{28}$$

where $2q$ is the number adjacent subcarriers which contribute the ICI effect significantly. Since, the ICI effects are considered from for $2q$ neighboring terms, the complexity is $O(2qN)$.

3.2 OIDF–PIC-DZFD Method

The working principle of OIDF–PIC-DZFD method is similar to DZFD–PIC-DZFD method. In OIDF–PIC-DZFD, OIDF method is used instead of DZFD method to estimate the initial rough signal in the initial stage.

3.3 OIDF–PIC-OIDF Method

This method is almost similar to the previous proposed cancellation methods. The first and second stage is exactly same as OIDF-PIC-DZFD method. The third stage is performed by adopting the iterative OIDF method and is given as follows. At first, temporary detected signal is obtained by multiplying $H^{H}(k, k)$ with ICI free signal $Y_{offICI}^{I}(k)$ as given below

$$Z_{offICI}^{I}(k) = H^{H}(k, k)Y_{offICI}^{I}(k) = G(k)\hat{X}^{I}(k) + W'(k) \tag{29}$$

$$Z_{off \, ICI}^{I}(k) = \left[Z_{(1)off \, ICI}^{I}(k) \left(Z_{(2)off \, ICI}^{I}(k) \right) \right]^{T} \quad W'(k) = \left[W_{(1)}'(k) \, W_{(2)}'(k) \right]^{T} \tag{30}$$

The OIDF method is applied to ICI free signal and is illustrated below.
[*Step* 1]: Set the variables as per higher diversity gain

If $\alpha_1(k) \geq \alpha_2(k)$, put variable $a = 1$ and $b = 2$

If $\alpha_1(k) < \alpha_2(k)$, put variable $a = 2$ and $b = 1$

[*Step* 2]: Detect initial data signal using DZFD method as given in (24) with respect to higher diversity gain

$$\hat{X}_a^{I}(k) = Q\left(R_{(a), DZFD}^{I}(k) \right) = X_a^{I}(k) + W_a'^{I}(k) \tag{31}$$

[*Step* 3]: Detects the final data signal through iteratively cancel the CCI effects

$$for \, I = 1 : P$$
$$\hat{X}_b(k) = Q\{(Z_{b, offICI}^{I}(k) - \beta^* \hat{X}_a^{I}(k)/\rho_b(k)\} \tag{32}$$

$$\hat{X}_a(k) = Q\{(Z^I_{a,offICI}(k) - \beta\hat{X}^I_b(k)/\rho_a(k)\}$$

$$\text{end}$$

(33)

This method gives better performance than its counterpart OIDF-PIC-DZFD method as it has additional DF operation.

Table 1. Computational Complexity

Method	Multiplication
TDBLF	$3 \times (2N)^3 + (2N)^2$
SDFSE	$\left[4 \times (C)^{2(2q+1)}\right]N$
DZFD-PIC-DZFD	$[10 + (8 + 4 \times 2q)I]N$
OIDF- PIC-DZFD	$[22 + (8 + 4 \times 2q)I]N$
OIDF- PIC-OIDF	$[22 + (14 + 4 \times 2q)I]N$

4 Computational Complexity

The TDBLF was proposed in [18] and is based on MMSE filter design and filtering procedure to cancel the ICI effects and needs total $3 \times (2N)^3 + (2N)^2$ complex multiplications. The SDFSE was proposed in [20]. It needs $4 \times (C)^{2(2q+1)}$ states where C denotes the constellation size and q is number of nearest subcarrier adjacent to the main diagonal. The proposed DZFD-PIC-DZFD method uses DZFD method to estimate the initial data which requires 10 complex multiplications [9]. In the second stage ICI free received signal requires $4 \times 2q$ complex multiplications. In the third stage, the ICI free received signal uses one tap zero forcing equalization which requires 8 complex multiplications. Hence total $10 + (8 + 4 \times 2q)I$ multiplications are needed for DZFD-PIC-DZFD method. OIDF-PIC-DZFD method is similar to DZFD-PIC-DZFD method instead it uses OIDF method to estimate the initial data symbol. In OIDF detection method, DZFD requires 10 complex multiplications for initial estimates of data signal, 4 complex multiplications require for $H^H Y$ operations and 6 complex multiplications for $\beta, \alpha_1, \alpha_2$. The first step of OIDF method involves no complex operation. Step 2 adopts DZFD method and therefore needs 10 complex operations. Step (3) and (4) needs two complex multiplication operations for $\beta^*(k)\hat{X}_a(k)$ and $\beta(k)\hat{X}_b(k)$. Hence, OIDF method needs total 22 complex multiplications. Total number of complex multiplications of DZFD-PIC-DZFD requires $22 + (8 + 4 \times 2q)I$. The first and second stage of OIDF-PIC-OIDF is similar to the previously discussed OIDF-PIC-DZFD method and thus needs $22 + (4 \times 2q)$ complex multiplication. In the third stage, initial data symbol is estimated using ZF method which requires 8 complex multiplications. $Z^I_{offICI}(k)$ requires 4 complex multiplications. The value of β and α are calculated in the first stage and is stored in a buffer and hence needs no extra complexity. Step (3) and (4) involves two complex multiplication operations i.e., $\beta^*(k)\hat{X}_a^I(k)$ and $\beta(k)\hat{X}_b^I(k)$. Hence, total $22 + (14 + 4 \times 2q)I$ complex multiplications are needed. The computation complexity of these interference cancellation methods is given in the Table 1.

Table 2. Simulation Parameters

Parameter	Value
Size of FFT	128
Number OFDM subcarriers	128
Size of CP	16
Frequency of carrier	2.5 GHz
Bandwidth of OFDM	1 MHz
Type of modulation	QPSK
Channel model	Exponential decaying PDP
Channel delay spread (d)	3
Total number of multipaths	8
Velocity of mobile	200/400 km/h
Channel Doppler spread ($f_d N T_s$)	0.06/0.12

5 Simulation Results

This section compares the conventional and proposed methods performance with respect to symbol error rate (SER) for several normalized Doppler frequency $f_d N T_s$. In this work, we have modelled the channel using exponential decaying power delay profile (PDP) [28, 29]. The l-th path power is given by $\sigma_l^2 = \sigma_0^2 \lambda^l$, $l = 0, 1, 2, ...L$. Power of the first path is $\sigma_0^2 = 1 - e^{-1/d} / 1 - e^{-(L+1)/d}$ and $d = \frac{-\tau_{rms}}{T_s}$ is the channel normalized delay spread. The symbol τ_{rms} denotes the rms delay spread of channel. The parameter $T_s = 1/W$ is the sampling time and W is bandwidth of OFDM system. Total number of multipaths (L) is calculated as $L = \tau_{max}/T_s$. $\tau_{max} = -\tau_{rms} ln A$, τ_{max} is the maximum delay of the chaanel. The parameter A is the ratio between the power of first path to the power of non-negligible path. For the parameters $d = 3$ and $A = -15$ dB, then number of multipaths is calculated as 8. Additionally, each multipath channel exhibits time variation and is modelled using Jakes sum-of-sinusoidal (SOS) [30]. The frequency of the carrier (f_c) is assumed as 2.5 GHz and bandwidth (W) is taken as 1 MHz. For mobile velocity of 200 and 400 km/h, the channel normalized Doppler spread (fdNTs) value becomes 0.06 and 0.12 respectively. Table 2 lists the total parameters and values used for simulation.

Figure 2 shows the performance of various CCI cancellation techniques for $f_d NT_s = 0.06$. The simulation result demonstrate that the Alamouti technique suffers performance degradation because of the occurance of both CCI and ICI effects. Although SIC technique achieves better result than the Alamouti method, its accuracy is limited to error propagation issue. The DZFD method outperforms SIC method, but as the Doppler frequency rises, its diversity gain declines. The DFD method produces superior results compared to DZFD. The List-SIC provides 10 significantly better result as compared to DFD method, but its complexity increases for higher order modulation. The OIDF method performs close to ML method. Following are the results of several CCI signal detection techniques listed in descending order: ML, OIDF, List-SIC, DF, DZFD, SIC and Alamouti.

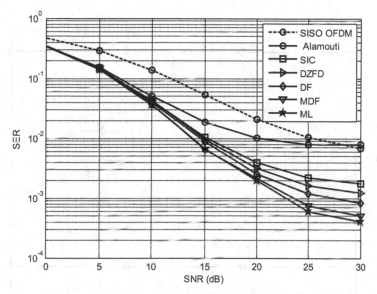

Fig. 2. Performance comparison of several CCI cancellation techniques for normalized Doppler spread $f_d NT_s = 0.06$

Figure 3 depicts the performance of DFD, ML and OIDF methods with various iterations (P) for $f_d NT_s = 0.06$ at 25 dB SNR. The result shows, the OIDF achieves significant performance than DFD method and is near to the ML method. It is demonstrated that the OIDF method achieves its optimal value with $P = 2$ iterations. However, it is seen from the simulation results that only cancelling the CCI effects does not provide sufficient performance.

The SER performance for various conventional which includes Alamouti, DZFD, OIDF, SDSFE, TDBLF and three proposed methods as shown in the Fig. 4 for $f_d NT_s = 0.6$. The result shows that the proposed DZFD-PIC-DZFD and OIDF-PIC-OIDF gives better performance than SDFSE method with q = 0. However, SDFSE with q = 2 outperforms the proposed DZFD-PIC-DZFD and OIDF-PIC-DZFD methods. The OIDF-PIC-DZFD gives better results than DZFD-PIC-DZFD method as it uses OIDF method instead of ZF to initial estimate the signal in the stage 1. The OIDF-PIC-OIDF gives better performance than its counterparts OIDF-PIC-DZFD method as it has additional OIDF operation but its computational complexity is slightly higher. From the Fig. 4, it is seen that the OIDF-PIC-OIDF method gives much better performance than TDBLF with much lower complexity as indicated in Table 1.

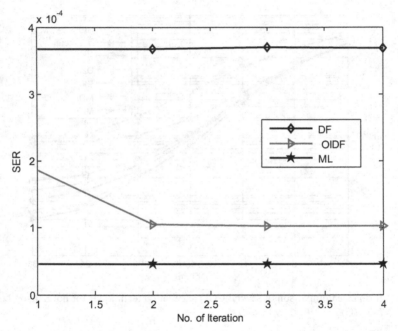

Fig. 3. SER vs no. of iteration of DF, ML and OIDF techniques for normalized Doppler spread $f_d NT_s = 0.06$

Figure 5 depicts the performance comparison of various signal detection methods for $f_d NT_s = 0.12$. It is observed that, TDBLF gives much better performance than OIDF-PIC-DZFD and DZFD-PIC-DZFD but lower as comparison to OIDF-PIC-OIDF method. The performance of several methods in ascending order are Alamouti, DZFD, DZFD-PIC-DZFD, OIDF, OIDF-PIC-DZFD, SDFSE, TDBLF and OIDF-PIC-OIDF. The BER vs normalized doppler spread performance of various signal detection methods is shown in Fig. 6. From the result, it is obvious that performance of various detection methods decreases with increases in normalized Doppler spread. The OIDF-PIC-OIDF method outperforms the SDFSE and TDBLF method irrespective of any mobile speed.

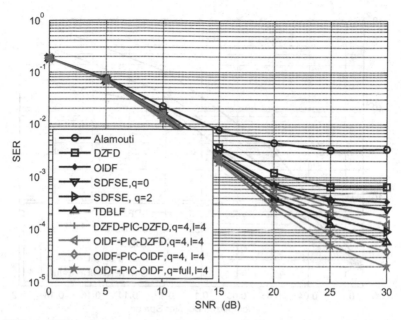

Fig.4. BER vs SNR performance comparison of several signal detection methods for $f_d N T_s = 0.06$

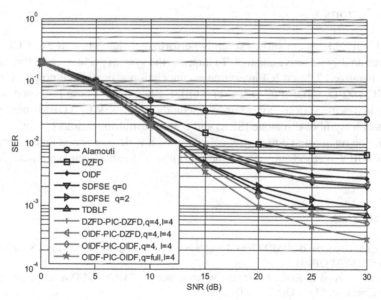

Fig.5. BER vs SNR performance comparison of several signal detection methods for $f_d N T_s = 0.12$

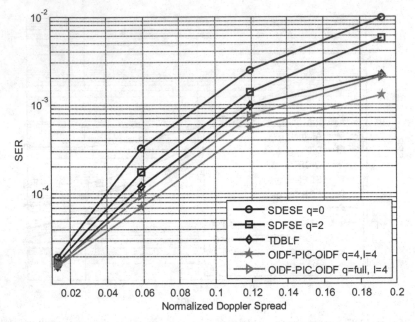

Fig. 6. SER vs normalized Doppler spread of various signal detection methods

6 Conclusions

The STBC-OFDM system suffres severely performance degradion due to CCI and ICI effects over doubly selective channel. To improve the system performance, three simplified joint iterative CCI and ICI interference cancelation schemes are proposed namely DZFD-PIC-DZFD, OIDF-PIC-DZFD and OIDF-PIC-OIDF. These proposed methods are compared with various conventional methods such as ML, TDBLF and SDFSE with respect to symbol-error-rate (SER) and complex multiplications operations. From the simulation results and complexity calculations, it is demonstrated that the proposed OIDF-PIC-OIDF method outperforms the conventional methods with significantly lower complexity.

References

1. Stuber, G.: Broadband MIMO-OFDM wireless communications. IEEE Trans. Signal Process. **92**(2), 271–294 (2004)
2. Yang, H.: A road to future broadband wireless access: MIMO-OFDM based air interface. IEEE Commun. Mag. **43**(1), 53–60 (2005)
3. Alamouti, S.M.: A simple transmitter diversity scheme for wireless communications. IEEE J. Sel. Areas Commun. **16**(8), 1451–1458 (1998)
4. Pham, Q.-V., Nguyen, N.T., Huynh-The, T., Le, L.B., Lee, K., Hwang, W.-J.: Intelligent radio signal processing: a survey. IEEE Access **9**, 83818–83850 (2021)
5. Djordjevic, I.B.: Advanced Optical and Wireless Communications Systems. Springer, Heidelberg (2017). https://doi.org/10.1007/978-3-319-63151-6

6. Lee, K.F., Williams, D.B.: A space-time coded transmitter diversity technique for frequency selective fading channels. In: IEEE Sensor Array and Multichannel Signal Processing Workshop, pp. 149–152 (2000)

7. Wee, J.W., Seo, J.W., Lee, K.T., Lee, Y.S., Jeon, W.G.: Successive interference cancellation for STBC-OFDM systems in a fast-fading channel. In: 61st IEEE Vehicular Technology Conference (VTC Spring), pp. 841–844 (2005)

8. Tso, C.-Y., Wu, J.-M., Ting, P.-A.: Iterative interference cancellation for STBC-OFDM systems in fast fading channels. In: Proceeding of IEEE Global Telecommunications Conference, GLOBECOM, pp. 1–5 (2009)

9. Li, C.M., Tang, I.T., Li, G.W.: Performance comparison of the STBC OFDM decoders in a fast-fading channel. J. Mar. Sci. Technol. 20(5), 534–540 (2012)

10. Lin, D.B., Chiang, P.H., Li, H.J.: Performance analysis of two branch transmit diversity block-coded OFDM systems in time-varying multipath Rayleigh-fading channels. IEEE Trans. Veh. Technol. 54(1), 136–148 (2005)

11. Cortez, J., et al.: A very low complexity near ML detector based on QRD-M algorithm for STBC-VBLAST architecture. In: 7th IEEE Latin-American Conference on Communications (LATINCOM), pp. 1–5 (2015)

12. Tu, H.-H., Lee, C.-W., Lai, I. W.: Low-complexity maximum likelihood (ML) decoder for space-time block coded spatial permutation modulation (STBC-SPM). In: 2019 International Symposium on Intelligent Signal Processing and Communication Systems (ISPACS), pp. 1–2 (2019)

13. Akhondi, M., Alirezapouri, M.A.: A modified ZF algorithm for signal detection in an underwater MIMO STBC-OFDM acoustic communication system. Ann. Telecommun. 78, 1–17 (2023)

14. Patra, J.P., Pradhan, B.B., Prasad, S.T., Singh, P.: An efficient signal detection technique for STBC-OFDM in fast fading channel. In: Chauhey, N., Thampi, S M., Jhanjhi, N.Z., Parikh, S., Amin, K. (eds.) Computing Science, Communication and Security. COMS2 2023. Communications in Computer and Information Science, vol. 1861, pp. 1–14. Springer, Cham (2023). https://doi.org/10.1007/978-3-031-40564-8_1

15. Russell, M., Stuber, G.L.: Interchannel interference analysis of OFDM in a mobile environment. In: Proceeding of IEEE 45th Vehicular Technology Conference, VTC, pp. 820–824 (1995)

16. Jeon, W.G., Chang, K.H.: An equalization technique for orthogonal frequency-division multiplexing systems in time-variant multipath channels. IEEE Trans. Commun. 47(1), 27–32 (1999)

17. Vlachos, E., Lalos, A.S., Berberidis, K.: Low-complexity OSIC equalization for OFDM-based vehicular communications. IEEE Trans. Veh. Technol. 66(5), 3765–3776 (2017)

18. Stamoulis, A., Diggavi, S.N., Al-Dhahir, N.: Intercarrier interference in MIMO OFDM. IEEE Trans. Signal Process. 50(10), 2451–2464 (2002)

19. Kim, J., Heath, R.W., Powers, E.J.: Reduced complexity signal detection for OFDM systems with transmit diversity. J. Commun. Netw. 9(1), 75–83 (2007)

20. Kim, J., Heath, R.W., Jr., Powers, E.J.: Receiver designs for Alamouti coded OFDM systems in fast fading channels. IEEE Trans. Wirel. Commun. 4(2), 550–559 (2005)

21. Dogan, H.: On detection in MIMO-OFDM systems over highly mobile wireless channels. Wirel. Pers. Commun. 86(2), 683–704 (2016)

22. Patra, J.P., Singh, P.: Efficient signal detection methods for high mobility OFDM system with transmit diversity. Arab. J. Sci. Eng. 44, 1769–1778 (2019)

23. Patra, J.P., Singh, P.: Joint iterative CCI and ICI cancellation for STBC-OFDM system in fast fading channel. In: 5th IEEE International Conference on Computer and Communication Technology, pp. 205–210 (2014)

24. Sheu, C.R., Liu, J.W., Huang, C.Y., Huang, C.C.: A low complexity interference suppression scheme for high mobility STBC-OFDM systems. In: IEEE Vehicular Technology Conference (VTC Spring) pp. 1–5 (2012)

25. Chen, H.C., Chang, W.K., Jou, S.J.: Low-overhead interference canceller for high mobility STBC OFDM systems. IEEE Trans. Circ. Syst. **60**(10), 2763–2773 (2013)

26. Soares, J.A., Mayer, K.S., Arantes, D.S.: Semi-supervised ML-based joint channel estimation and decoding for m-MIMO with Gaussian inference learning. IEEE Wirel. Commun. Lett. **12**, 2123–2127 (2023)

27. Marey, M., Mostafa, H., Alshebeili, S.A., Dobre, O.A.: STBC recognition for OFDM transmissions: channel decoder aided algorithm. IEEE Commun. Lett. **26**(7), 1658–1662 (2022)

28. Cho, Y.S., Kim, J., Yang, W.Y., Kang, C.G.: MIMO-OFDM Wireless Communications with MATLAB. Wiley, Singapore (2010)

29. Stüber, G.L.: Principles of Mobile Communication. Kluwer, London (2001)

30. Jakes, W.C.: Microwave Mobile Channels. Wiley, New York (1974)

Managing Multiple Identities of IoT Devices Using Blockchain

Shachi Sharma$^{(\boxtimes)}$ (iD), Santanu Mondal, and Shaheen Ishrat (iD)

Department of Computer Science, South Asian University, New Delhi 110068, India
shachi@sau.int

Abstract. The popularity of IoT has opened avenues for many new solutions such as social network of devices and web of things where IoT devices may have multiple logical identities. Managing multiple logical identities of voluminous number of IoT devices securely is a challenging task. As a result, efficient identity management of IoT devices has become an important research topic. The existing literature on identity management of IoT devices is currently restricted to only one logical identity per device. The paper presents a solution using blockchain for managing multiple identities of IoT devices. The usage of blockchain also helps in authentication of mobile IoT devices using logical identities. The implementation of the solution is carried out using Hyperledger blockchain platform and its performance is evaluated using a simulator developed in Python. The results validate the feasibility and efficiency of the proposed solution.

Keywords: Blockchain · Internet of Things · Multiple identities · Performance analysis

1 Introduction

In recent years, a voluminous growth in number of connected devices has been observed [18] and this pattern is predicted to continue for next one decade [16]. This growth is attributed to the wider adoption of Internet of Things (IoT) which is envisioned to change human life significantly by enabling machines and devices to interact with physical world. Because of this, new variants of networking in the form of Web of Things (WoT) and Social Internet of Things (SIoT) have emerged [17,21]. The WoT provides scalable and interoperable framework for creating smart applications and systems by merging many diverse IoT platforms with World Wide Web (WWW) [21]. The paradigm of SIoT allows distinct devices to create their own social network to achieve better performance, functionality and efficiency [17]. The devices in these solutions are not characterized by their physical identifiers but logical identities. Logical identity refers to a set of device's attributes that can be virtual or real [22].

Identity management is the process of securely storing and retrieving identities of devices [16]. As highlighted by Sharma *et al.* [13], *"Searching a device in*

© The Author(s), under exclusive license to Springer Nature Switzerland AG 2024
K. K. Patel et al. (Eds.): icSoftComp 2023, CCIS 2031, pp. 137–147, 2024.
https://doi.org/10.1007/978-3-031-53728-8_11

heterogeneous IoT environment requires one to either know the physical identity of the device or domain name of the organization owing the device. Other than this, there are instances when devices are to be searched based on criterion like location, functionality, owners i.e. the attributes characterizing devices, identity". The logical identities play an important role in identifying devices in WoT and SIoT, like URL in WWW or identity of humans in social networks.

The traditional IoT identity management solutions adopt centralized architecture where a central authority serves as the gatekeeper for management, authentication and access control of devices' identity. These systems use many protocols, such as Transport Layer Security (TLS) and X.509, to establish secure communication between IoT devices and centralized server. However, this architecture is vulnerable to single point of failure and susceptible to attack. The whole IoT ecosystem gets compromised if the centralized authority is attacked. Also, the centralized architectures fail to scale for supporting increasing number of IoT devices. The number and speed of devices registration and authentication requests lead to delays in centralized systems. To overcome the challenges of centralized identity management solutions in IoT, distributed architectures have also been proposed in the literature. Distributed identity management is a complex task, thus requires careful planning and implementation to address the unique challenges posed by the heterogeneous nature of IoT ecosystems. A promising technology adopted for distributed identity management is blockchain. The Blockchain is a peer to peer networking technology that can be used to maintain a tamper-proof record of device identities and transactions. This helps in ensuring security and trustworthiness of identity management process. A light weighted architecture for blockchain based identity management has been proposed by Bouras et al. [7] in which management functions and life cycle of device's identity in the network are discussed in detail. A comprehensive survey of many blockchain based identity management approaches has been provided by Lo et al. [2]. A major drawback of these existing works lies in their limitation to support either physical device identifiers or one logical identity per device. The new technologies like WoT and SIoT require systems that are capable of managing multiple identities of a device in dynamic environment of IoT. The paper makes an attempt towards fulfilling this research gap.

A method for managing multiple identities of IoT devices using blockchain is proposed in the paper. The paper is organized into five sections. A summary of related work on identity management systems is provided in Sect. 2. A blockchain based method, called MIDMID, is proposed in Sect. 3 for multiple identities management. The feasibility and performance analysis of the rudimentary implementation of the proposed method is present in Sect. 4. The last Sect. 5 contains conclusion and future work.

2   Related Work

The area of identity management is not new. It came into existence with Internet. In order to access a service on Internet, any user first needs to register by

Logical identity

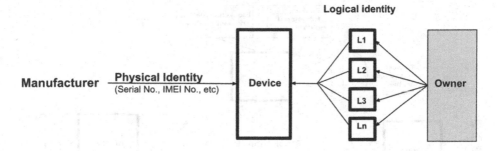

Fig. 1. Physical identity versus multiple logical identities of a device.

creating a digital identity. With proliferation of services, users started to create multiple digital identities for multiple services. This led to the development of efficient techniques for managing digital identities of users in separate specialized and centralized domain specific systems [1]. Followed are the federated identity management systems [10] that builds trust relationships between identity providers there by allowing users in one secure domain to access services from another domain. In the context of IoT devices, the need for logical identifier was discussed in a decade old work by Sharma *et al.* [13]. The usage of logical identifiers in device as well as service discovery has also been explored [13,22,23]. Most of the earlier approaches of centralized identity management cannot be applied in IoT because of its characteristic requirements like scalability, interoperability, mobility and dynamic environment [24]. The distributed and decentralized solutions are found to be suitable for addressing IoT specific requirements as they scale well. However, synchronization is more difficult in distributed approaches [14], when large number of devices need to be managed. Hence, blockchain based identity management is gaining popularity for IoT devices.

Blockchain by design is an immutable distributed ledger with decentralized control where transactions are stored. Its application in insecure IoT environment is promising. Hence, many blockhain based identity management systems have been proposed in the literature. A notable solution has been proposed by Bassam [1] in which public Ethereum blockchain-based identity system is developed using web-of-trust model and smart contracts. This system can be accessed by an entity (for example, a company) which can find attributes of another entity. Clearly, this system has not been built considering requirements of IoT environment. The readers can refer to [2–4,11,12,15] for more details of these types of identity management systems. A good survey of identity management systems considering IoT specific requirements has been carried out by Cremonezi *et al.* [9]. Kuperberg [14] reviewed the functional and non-functional requirements for blockchain based identity and access management by defining 75 criterion. Butun and Osterberg [8] highlighted various authorization, authentication and revocation methods that must be supported in a blockchain enabled identity management and access control system. This work suggests that permissioned blockchain platforms should be preferred for IoT due to the security and privacy

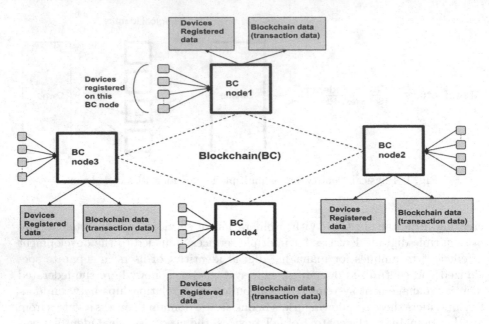

Fig. 2. Proposed solution for managing multiple identities of IoT devices using blockchain.

offered by it. In another work, Ren [20] proposed a blockchain based identity management and access control approach utilizing Bloom filter for edge computing environment. It is worth mentioning that edge computing is a paradigm to process data generated by IoT devices in real-time to fulfill the requirements of mission-critical applications such as smart grids or driverless cars. Nyante [19] utilized a blockchain gateway connecting blockchain with off chain regular world for creating a safe privacy maintaining framework. Bernabe *et al.* [6] designed a holistic and privacy-preserving IoT solution for identity management and authentication. A claims-based strategy was developed for authentication and access control. Bao *et al.* [5] proposed a three-tier architecture comprising of authentication, blockchain, and application layers for identity authentication, access control and privacy protection for IoT devices. A light weight consortium blockchain solution for IoT devices identity management was provided by Bouras [7]. A major drawback of all these existing identity management solution is that they allow to register one logical or physical identity of a device. The WoT and SIoT require to manage multiple logical identities of a device in future. This motivates us to undertake the research presented in this paper.

3   MIDMID: The Proposed Method

In this section, a new method called MIDMID (Multi-Identity Management of IoT Devices) using blockchain is proposed. The requirement for multiple logical

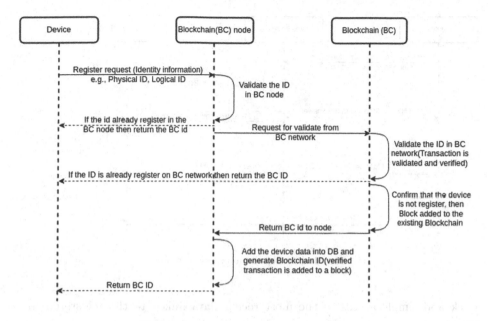

Fig. 3. Registration process of a device.

identities of IoT devices exists due to applications like SIoT and WoT as well as for better authentication, access control, data segmentation, interoperability and regulatory compliance. It is worth highlighting that a unique physical identity is assigned to a device by the manufacturer whereas the owner of the device can create multiple logical identities as illustrated in Fig. 1. These logical identities are assigned based on the devices specific characteristics, requirements, usage, and operating circumstances. The logical identities serve as digital representation of devices inside an ecosystem or network of its owners.

Blockchain provides a useful way to manage IoT devices in dynamic environment where movement of devices may occur between networks. The blockchain network is made up of several blockchain nodes called peers, each of which is responsible for maintaining the blockchain (a ledger of transactions). IoT devices connect with blockchain peers allowing them to communicate with the blockchain network. Since blockchain keeps record of each transaction occurring between devices, it provides a secure communication path. In Fig. 2, we propose a system architecture for managing multiple identities of IoT devices using blockchain network. Each blockchain node includes two key elements viz. ledger and database that stores data. The blockchain shared ledger is distributed and tamper-proof. It provides transparency and immutability by recording every transaction in a sequential and linked manner. All network nodes share this ledger ensuring that all participants have the same blockchain history. The second component of each node is a database that maintains the current state of the blockchain for a particular IoT device registered to the blockchain. The database enables to keep track of the current state and related information of IoT devices. It allows for

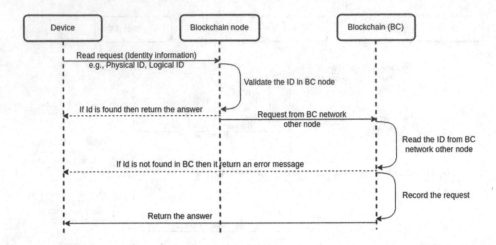

Fig. 4. Details of query processing.

quick and simple access to the most recent data linked to the registered IoT devices.

In a typical IoT identity management system, the CRUD (create, read, update, delete) operations are used to manage identities and related information. The system can add a new device to the blockchain identity management system using the "create" function. This is also called "register" in some implementations. The "read" function allows the system to retrieve device information from the system. It allows authorised and authenticated device identity. It is also referred as "query" in some systems. The "update" function in the system helps to modify the device identity. The "delete" or "deregister" ore "revoke" function helps to restrict or remove any identity. We present the details of CRUD operations in the proposed method

3.1   CRUD Operations

Create/Register Device: The sequence of steps for registering device is explained in Fig. 3. An IoT device initiates the process by submitting a request for registration through a blockchain node. The device provides identity information viz. physical identifier, one or more logical identifiers and other meta information. The blockchain node first searches its database to check if the physical identifier already exists or not. If the information is present in the node's database, it returns a message indicating that device has already been registered. If the physical identifier of the device is not present in the node's database, then blockchain node tries to confirm from other peers in the blockchain network. The node forwards the request to other nodes of the blockchain network, asking them to authenticate the device identification and check if it is registered with any other node. Each node on receiving the validation request searches in its database for the specified physical identifier. If the device is already registered

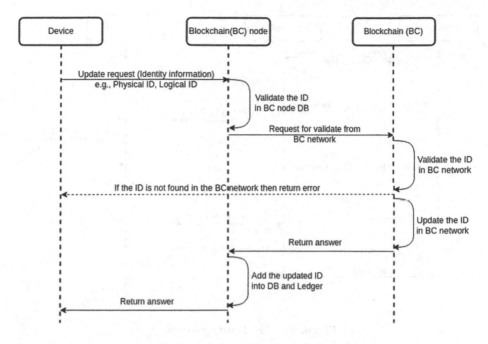

Fig. 5. Process of update device.

with any other node, that node responds to the requesting node with the device's blockchain ID (BC id) notifying that the device is already registered. If the device information is not available with any other peer nodes in the blockchain then first a new BC id is created for the device and a new transaction gets recorded in the blockchain. The node then updates the identity information in its database along with BC id. A success reply with BC id is sent back to the device.

Read/Query Device: The query procedure is initiated by the IoT device through a request to one of the blockchain node as illustrated in Fig. 4 including either the physical identifier or BC id or one of the logical identifier. After receiving the request, the blockchain node examines its own database first to find out if the specified identifier already exists. If the device is registered with the querying node then a message with list of logical identifiers and meta information is sent back to the device. If the device is not registered with the querying node then other peers are interrogated. The peers then try to locate device in their databases and respond back with appropriate answer viz. if device is registered then list of logical identifiers, BC id and meta information are returned or else an error is returned. All the sequence of exchanged messages gets recorded in blockchain ledger. The requesting device is accordingly updated with the details.

Update Device: A device can add a new logical identifier, remove an existing logical identifier or modify meta information through this operation. The

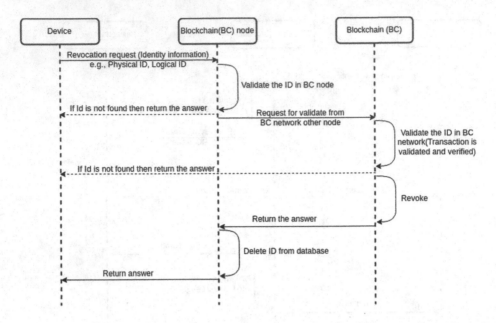

Fig. 6. Revoke identity process.

procedure of handling update request is a mix of registration and query process as depicted in Fig. 5. The device must include either physical identifier or BC id along with information it intends to modify in the update message. After receiving the request, the blockchain node examines its own database first and if the details of devices are not there then it communicate with other peers following same sequence of messages as in register and query processing. This allows a moving IoT device to update details from other blockchain peers as well.

Delete Device: As shown in Fig. 6, the procedure begins with the IoT device sending a revocation request to a blockchain node including wither the physical identifier or BC id. Again the same sequence of actions is followed as in case of registration and updation so that the details of the device gets removed from the database if it is available with any peer.

4   Performance Analysis

We have chosen Hyperledger fabric blockchain platform for validating the feasibility of the MIDMID method. The reason for this choice is because of the fact that Ethereum blockchain platform does not support off chain database where identities of devices can be stored. In contrast, Hyperledger support external databases like Couch DB and Level DB. We use Couch DB in the implementation. The smart contracts are written to implement the sequence diagrams

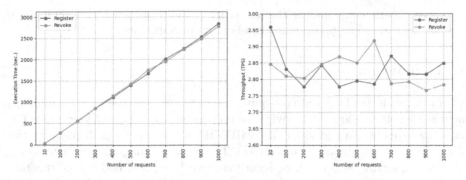

(a) Execution time of register and revoke requests.

(b) Throughput of register and revoke requests.

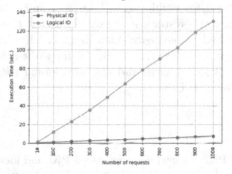

(c) Execution time of query requests.

Fig. 7. Results of performance analysis.

discussed in Sect. 3. The update operation has multiple variants like adding a new logical id to a device, renaming a logical id, deleting a logical id.

Experiments are run by creating a test bed comprising of two peers blockchain network using Hyperledger fabric version 2.5. A simulator is also coded in Python to emulate IoT devices. The configuration of test computers is Intel core i7-8700 3.20GHz CPU, 16 GB RAM, 64-bit operating system on x64-based processor. For performance monitoring of the blockchain network, Hyperledger explorer v1.1.8 has been used.

The results of experiments are presented in Figs. 7a–7c. As expected the execution time of both registration and revoke increases with number of requests. In contrast the throughput varies for these operations. The interesting behavior is observed with query operation. The execution time of queries with physical identifiers remains almost constant with number of devices while it rises linearly for logical identifiers. With physical identifiers specified, the device can be searched instantly in the database because of uniqueness. Query with logical identifiers requires more search in the database resulting in high execution time.

These experiments validate the feasibility of the proposed MIDMID method and also provide some preliminary inside in its performance.

5 Conclusion

Advancement in WoT and SIoT requires a secure and scalable identity management system that allows IoT devices to register multiple logical identities. The paper has presented a method, called MIDMID, for managing multiple identities of IoT devices using blockchain platform. The sequence diagrams for registering, updating, querying and revoking identities have been designed. A rudimentary implementation of the proposed method on Hyperledger blockchain has been done by developing smart contracts in Javascript. A simulator in Python language has also been built for emulating IoT devices and is used in performance analysis of the rudimentary system. It is examined that the search time is high when a device is searched using logical identifier. The implementation confirms the feasibility of the proposed method. The future work aims to test the MID-MID method using real IoT devices and also developing the solution for public blockchain platform such as Ethereum.

References

1. Al-Bassam, M.: SCPKI: a smart contract-based PKI and identity system. In: Proceedings of the ACM Workshop on Blockchain, Cryptocurrencies and Contracts, pp. 35–40 (2017)
2. Augot, D., Chabanne, H., Chenevier, T., George, W., Lambert, L.: A user-centric system for verified identities on the bitcoin blockchain. In: Garcia-Alfaro, J., Navarro-Arribas, G., Hartenstein, H., Herrera-Joancomarti, J. (eds.) Data Privacy Management, Cryptocurrencies and Blockchain Technology. Lecture Notes in Computer Science(), vol. 10436, pp. 390–407. Springer, Cham (2017). https://doi.org/10.1007/978-3-319-67816-0_22
3. Augot, D., Chabanne, H., Clémot, O., George, W.: Transforming face-to-face identity proofing into anonymous digital identity using the bitcoin blockchain. In: 2017 15th Annual Conference on Privacy, Security and Trust (PST), pp. 25–2509. IEEE (2017)
4. Axon, L.: Privacy-awareness in blockchain-based PKI. Cdt Tech. Pap. Ser. **21**, 15 (2015)
5. Bao, Z., Shi, W., He, D., Chood, K.K.R.: IoTChain: a three-tier blockchain-based IoT security architecture. arXiv preprint: arXiv:1806.02008 (2018)
6. Bernal Bernabe, J., Hernandez-Ramos, J.L., Skarmeta Gomez, A.F.: Holistic privacy-preserving identity management system for the internet of things. Mob. Inf. Syst. **2017** (2017)
7. Bouras, M.A., Lu, Q., Dhelim, S., Ning, H.: A lightweight blockchain-based IoT identity management approach. Future Internet **13**(2), 24 (2021)
8. Butun, I., Österberg, P.: A review of distributed access control for blockchain systems towards securing the internet of things. IEEE Access **9**, 5428–5441 (2020)
9. Cremonezi, B., Vieira, A., Nacif, J.A., Nogueira, M.: Survey on identity and access management for internet of things (2020)

10. Gomes, P., Cavalcante, E., Rodrigues, T., Batista, T., Delicato, F.C., Pires, P.F.: A federated discovery service for the internet of things. In: Proceedings of the 2nd Workshop on Middleware for Context-Aware Applications in the IoT, pp. 25–30 (2015)

11. Halpin, H.: NEXTLEAP: decentralizing identity with privacy for secure messaging. In: Proceedings of the 12th International Conference on Availability, Reliability and Security, pp. 1–10 (2017)

12. Hardjono, T., Pentland, A.: Verifiable anonymous identities and access control in permissioned blockchains. arXiv preprint: arXiv:1903.04584 (2019)

13. Kapoor, S., Sharma, S., Srinivasan, B.R.: Attribute-based identification schemes for objects in internet of things (2013), uS Patent 8,495,072

14. Kuperberg, M.: Blockchain-based identity management: a survey from the enterprise and ecosystem perspective. IEEE Trans. Eng. Manage. **67**(4), 1008–1027 (2019)

15. Liu, Y., Zhao, Z., Guo, G., Wang, X., Tan, Z., Wang, S.: An identity management system based on blockchain. In: 2017 15th Annual Conference on Privacy, Security and Trust (PST), pp. 44–4409. IEEE (2017)

16. Nakamoto, S.: Bitcoin: a peer-to-peer electronic cash system. Decentralized business review, p. 21260 (2008)

17. Nitti, M., Atzori, L., Cvijikj, I.P.: Network navigability in the social internet of things. In: 2014 IEEE World Forum on Internet of Things (WF-IoT), pp. 405–410. IEEE (2014)

18. Nofer, M., Gomber, P., Hinz, O., Schiereck, D.: Blockchain. Bus. Inf. Syst. Eng. **59**, 183–187 (2017)

19. Nyante, K.: Secure identity management on the blockchain. Master's thesis, University of Twente (2018)

20. Ren, Y., Zhu, F., Qi, J., Wang, J., Sangaiah, A.K.. Identity management and access control based on blockchain under edge computing for the industrial internet of things. Appl. Sci. **9**(10), 2058 (2019)

21. Sciullo, L., Gigli, L., Montori, F., Trotta, A., Felice, M.D.: A survey on the web of things. IEEE Access **10**, 47570–47596 (2022). https://doi.org/10.1109/ACCESS.2022.3171575

22. Sharma, S.: Attribute based discovery architecture for devices in internet of things (IoT). In: 2019 IEEE 5th International Conference for Convergence in Technology (I2CT), pp. 1–4 (2019). https://doi.org/10.1109/I2CT45611.2019.9033565

23. Sharma, S., Kapoor, S., Srinivasan, B.R., Narula, M.S.: HICHO: attributes based classification of ubiquitous devices. In: Puiatti, A., Gu, T. (eds.) Mobile and Ubiquitous Systems: Computing, Networking, and Services, pp. 113–125. Springer, Berlin Heidelberg, Berlin, Heidelberg (2012). https://doi.org/10.1007/978-3-642-30973-1_10

24. Zhu, X., Badr, Y.: A survey on blockchain-based identity management systems for the internet of things. In: 2018 IEEE International Conference on Internet of Things (iThings) and IEEE Green Computing and Communications (GreenCom) and IEEE Cyber, Physical and Social Computing (CPSCom) and IEEE Smart Data (SmartData), pp. 1568–1573. IEEE (2018)

Machine Learning Enabled Image Classification Using K-Nearest Neighbour and Learning Vector Quantization

J. E. T. Akinsola[1]([∅]) [iD], F. O. Onipede[1] [iD], E. A. Olajubu[2] [iD],
and G. A. Aderounmu[2] [iD]

[1] First Technical University, Ibadan, Nigeria
akinsolajet@gmail.com
[2] Obafemi Awolowo University, Ile-Ife, Nigeria

Abstract. Nowadays, the classification of images is used to bridge the gap between human vision as well as computer vision to identify images by machines in the same way humans do. It concerns assigning the appropriate class for a provided image. The major problems encountered in the classification of images is the representation of image vector as well as image feature extraction. To overcome major image classification issues, machine learning algorithms such as K-NN and LVQ are implemented using Scikit-learn and Python packages on the Iris dataset. LVQ and KNN models were developed using a 70:30 split ratio for training and testing the models. The algorithm's performance was compared using machine learning performance metrics such as Accuracy, F1 Score, Recall, and Precision. According to the findings, KNN is a superior option for classification tasks that demand high accuracy-based parameter setting, and the implementation followed specifically due to the metrics result with 96.67% accuracy, 1.00 precision, 0.89 recall, and 0.94 F1 Score. Hence, KNN can be applied for effective image classification problems. KNN and LVQ both have their strength and weaknesses depending on the problem at hand. The study, therefore, recommends the use of other machine learning algorithms such as random forest, and decision tree with hybridization of ensemble learning on the same dataset for optimal comparison.

Keywords: Artificial Intelligence · Data Normalization · Deep Learning · Image Classification · KNN · Learning Vector Quantization · Machine Learning

1 Introduction

Nowadays, the classification of images is used to bridge the gap between human vision as well as computer vision to identify images by machines in the same way humans do. It concerns assigning the appropriate class for a provided image [1]. The major problem encountered in image classification is representation of image vector as well as image feature extraction [2]. Classification of an image is the way of allocating pixels in a digital image into interest classes. The goal of classification of image is the identification of images features that is unique. To carry out classification on a set of data into several

K. K. Patel et al. (Eds.): icSoftComp 2023, CCIS 2031, pp. 148–163, 2024.
https://doi.org/10.1007/978-3-031-53728-8_12

categories or classes, the association between the classes and the data into which they are categorized must be well known [1]. Classification task is a well-known task in the community of machine learning (ML) as well as deep learning [3]. Flowers are challenging to categorize [4] due to their close similarities. With the help of machine learning algorithms such as Learning Vector Quantization and K-Nearest Neighbor, the process of classification of flowers is made easier with high rate of efficiency. Flower image classification also plays an important role of sustaining the ecological balance in plants.

The domain of artificial intelligence (AI) known as "machine learning" is concerned with the development of statistical models and algorithms that enable computers to "learn" from data and predict the future without being explicitly programmed. AI simulates human cognitive processes and behavior using machines. It has a wide range of academic disciplines, including psychology, linguistics, and philosophy [5]. The basic goal of ML is to create models that can identify patterns in data and produce predictions. To do this, a model is trained on a sizable dataset and its parameters are tuned to enable precise prediction. The field of ML focuses on assisting computing systems to learn from data on how to automatically perform required operations [6]. Predictive analytics systems that are based on ML are widely used in different areas. There is now presence of automations that are of high level with great power of prediction using reinforcement learning as well as deep learning techniques [7].

Among the several kinds of machine learning are supervised learning, unsupervised learning, semi-supervised learning, and reinforcement learning. The selection of which type of learning to use depends on the particular use cases and applications that each type of learning provides. ML uses different classification algorithms for performance metrics evaluation which belong to six major classes or categories which are artificial neural network (ANN) classifiers, Bayes classifiers, tree classifiers, function classifiers sequential minimal optimization and also lazy classifiers [8].

Therefore, this research has contributed to the body of knowledge as follows:

i. building of machine learning model using Iris dataset on different splitting ratios.
ii. improved performance result than the existing models.
iii. comparative analysis to choose the best image classification model.
iv. applications of KNN and LVQ mathematical models for image classification

This study is divided into five sections which are introduction, related work, materials and methods, results and discussion, and lastly conclusion.

2 Related Works

Different KNN and LVQ algorithms approaches are elucidated accordingly.

2.1 K-Nearest Neighbors (KNN)

The K-Nearest Neighbors (KNN) machine learning method, is employed for classification and regression applications [9]. KNN detects the k closest neighbors to a new data point in the training data and uses them to forecast the class or value of the new data point. The user-defined parameter k determines the number of neighbors [10], and the forecast is based on the average value or majority class of the k closest neighbors.

As an instance-based learning algorithm, KNN saves the data used for training and bases its predictions on those instances rather than utilizing a model built from the data used for training. KNN can handle correlations between characteristics and target variables that are not linear. However, can be computationally expensive and prone to overfitting, especially when the training data is abundant or the number of features is high. Due to its simplicity and convenience of use, KNN is still a frequently used technique, particularly for small and medium-sized datasets. Cross-validation or trial-and-error selection can be used to choose the user-defined value, k, which represents the number of closest neighbors.

2.1.1 How KNN Works

KNN mechanism works as described thus

i. Compute the spread between the new data point and every other data point in the dataset.
ii. k nearest data points should be chosen (that is, the k closest data points with the lowest distances)
iii. Using the majority class of the new data point's k nearest neighbours, determine the class of the new data point.

Two data points can be compared using any distance metric, including the Manhattan and Euclidean distances. KNN is a popular technique for classification and regression issues since it is straightforward yet effective. Outlier detection and dimensionality reduction are other uses. KNN is computationally expensive. Overall, KNN is a flexible and strong algorithm that may be used to solve a variety of machine learning issues.

2.1.2 The KNN Variations

The major KNN variations are given accordingly.

i. **Weighted KNN:** In this form, the target class or value prediction is established on a weighted average of the K nearest neighbours, with the weights being defined by the distances between the sample and its neighbours.
ii. **KNN with Dimensionality Reduction:** The author suggests using dimensionality reduction methods, such as PCA or LLE, before applying KNN to alleviate the problem of the "curse of dimensionality" [11].
iii. **KNN with Feature Selection:** According to the author, KNN can perform better when there are fewer duplicated or irrelevant features in the dataset. Examples of such strategies include mutual information and the chi-squared test [12].
iv. **KNN with Hybrid Distance Measures:** To enhance the performance, KNN uses hybrid distance measures, which incorporate various distance metrics.
v. **KNN for Regression:** This gives an overview of the use of KNN to regression issues where the target variable is continuous rather than categorical.

The distance between two instances is calculated based on their similarity in terms of patterns and trends. KNN is useful for time series data, where the sequence of observations is represented as a single instance. Different KNN methods and several of its variants and expansions are provided in this study with review of the techniques.

2.2 Learning Vector Quantization (LVQ)

Additionally, inspired by biological representations of brain systems. LVQ is a type of artificial neural network. Its neural network is trained using a competitive learning strategy akin to the Self Organizing Map as its base, which is a prototype supervised learning classification system. It addresses the multiclass classification problem. The two layers that comprise of LVQ are the input layer and the output layer [13]. The design of the LVQ is shown in Fig. 1 for any collection with a variety of classes and input features.

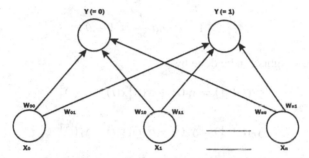

Fig. 1. Architecture of Learning Vector Quantization [13].

2.2.1 Learning Vector Quantization (LVQ) Variations

There are six major LVQ variation which are described thus.

i. **Learning Vector Quantization 1, or LVQ1:** This is the first LVQ technique that Kohonen proposed. It is often referred to as the "Winner Takes All" approach [14].
ii. **LVQ2:** This LVQ1 variation gives each characteristic in the input vector a weight, enabling more precise classifications [15].
iii. **LVQ3:** This upgrades from LVQ2 enables online learning and the real-time modification of the weight vectors.
iv. **ALVQ or adaptive LVQ:** this is a variant of LVQ that modifies learning rate in response to classifier performance.
v. **Self-Organizing Map (SOM):** this is an unsupervised learning technique frequently thought of as a subset of LVQ. Data visualization and dimensionality reduction use it.
vi. **Adaptive Resonance Theory (ART):** this is a form of neural network developed for pattern identification and classification and was influenced by LVQ.

2.2.2   Learning Vector Quantization (LVQ) Variations

There are three additional iterations of the LVQ algorithm: LVQ2, LVQ2.1, and LVQ3. They are Teuvo Kohonen's creation.

1. Algorithm LVQ2

The second improved iteration of the LVQ algorithm mimics Bayesian decision theory. Both LVQ2 and LVQ, have the same stages with some differences. In certain situations, the LVQ2 algorithm employs weights [15], such as:

a. when the classification of the input vector is wrong.
b. when the subsequent nearest vector is successfully classified.
c. when the decision boundary can be obtained using just the input vector.

Only when the input vector fits into a window that can be changed in this case does learning occur [15]. Equation 1 shows the learning rate.

$$\frac{d_c}{d_r} > 1 - \theta \ and \ \frac{d_r}{d_c} > 1 + \theta \tag{1}$$

Updating the weights can be done by:

$$y_c(t+1) = y_c(t) + \alpha(t)\big[x(t) - y_c(t)\big] \tag{2}$$

$$y_r(t+1) = y_r(t) + \alpha(t)\big[x(t) - y_r(t)\big] \tag{3}$$

where, x is the input vector, y_c is the winning vector, y_r is the other closet vector, d_c is the distance between x & y_c, d_r is the distance between x & y_r, θ total number of training samples and α is the learning rate.

2. LVQ2.1 Algorithm

A prevalent LVQ variation is LVQ2.1. In LVQ2, the weights are changed in two separate circumstances: first, when the winning vector has the same class label as the following vector, and second, when the winning vector has a different class label. In contrast, either vector may have the same class labels in LVQ2.1. The prerequisite for the window where the input vector fits is shown in Eqs. 4 and 5 [15].

$$min\left(\frac{d_{c1}}{d_{c2}}, \frac{d_{c2}}{d_{c1}}\right) > (1 - \theta) \tag{4}$$

$$max\left(\frac{d_{c1}}{d_{c2}}, \frac{d_{c2}}{d_{c1}}\right) < (1 + \theta) \tag{5}$$

The weights can be updated by:

$$y_{c1}(t+1) = y_{c1}(t) + \alpha(t)\big[x(t) - y_{c1}(t)\big] \tag{6}$$

$$y_{c2}(t+1) = y_{c2}(t) + \alpha(t)\big[x(t) - y_{c2}(t)\big] \tag{7}$$

3. LVQ3 Algorithm

When the input vector, the winning vector, and the next-closest vector all share the same class label, learning is further extended in LVQ3. The window's state here can be:

$$min\left(\frac{d_{c1}}{d_{c2}}, \frac{d_{c2}}{d_{c1}}\right) > (1 - \theta)(1 + \theta) \qquad (8)$$

The weights can be updated by:

$$y_{c1}(t + 1) = y_{c1}(t) + m\alpha(t)\big[x(t) - y_{c1}(t)\big] \qquad (9)$$

$$y_{c2}(t + 1) = y_{c2}(t) + m\alpha(t)\big[x(t) - y_{c2}(t)\big] \qquad (10)$$

where m is a balancing constant that has a range of 0.1 to 0.5.

2.2.3 Significance of LVQ

The importance of LVQ as it concerns image classification are:

1 **Automatic Classification:** LVQ is known to carryout classification of input vector automatically [16]. This aids quick identification of characteristics of images.
2 **Creation of Prototype:** LVQ makes it easier for experts to translate to respective application domain [17] through the creation of prototype.
3 **Multidimensional Data Processing:** LVQ can be used for multidimensional data processing on multidimensional data that has noisy inference [16].

3 Materials and Methods

This section discusses the various approaches of data mining as well as machine learning algorithms used for model building with various equations and how the equations are used to achieve their operational goals. The process, techniques as well as tools for the implantation are also keenly discussed.

3.1 Dataset Description

A machine learning dataset is a collection of data used to train a model. A dataset is used as an instance to teach the machine learning algorithm how to make predictions Shemir (2023). Ronald Fisher first proposed the Iris dataset in 1936 in his paper titled "The use of numerous measurements in taxonomic problems," and it has since grown to be a well-liked dataset for showcasing the power of machine learning techniques. In the ML world, the Iris dataset is commonly used to evaluate how KNN algorithm performs. Sepal length, sepal width, petal length, and petal width are the four traits that distinguish each of the 150 samples of Iris plants. The classification problem's goal is to identify the species of iris plant based on these four traits. The collection contains information on the Iris setosa, Iris virginica, and Iris versicolor species.

Implementation was done using Jupyter Notebook. The models were developed using Python programming and scikit-learn, pandas, NumPy, seaborn, matplotlib as well as performance metrics libraries. Table 1 gives the summary of dataset description.

Table 1. Dataset Description.

Features	Species	Iris Sample size
length of sepal	Iris setosa	150
width of sepal	Iris virginica	
length of petal	Iris versicolor	
width of petal		

3.1.1 Data Preparation Techniques Used in Iris Dataset

The following data preparation techniques were used in the course of this study.

i. **Data cleaning:** It is critical to identify any missing or erroneous values in the dataset and to treat them correctly. This can entail eliminating records with blank fields or imputing missing values.
ii. **Data normalization:** This helps ML algorithms perform better. The features in the Iris dataset have different sizes and units of measurement. Standard Scaler was used to normalize the Iris dataset to ensure the effective utilization of the dataset.
iii. **Data division:** The Iris dataset is split into training set and test set. The Iris dataset was split into 70% training and 30% testing for building the models.
iv. **Data transformation:** In some circumstances, it may be advantageous to alter the dataset's properties to improve their suitability for analysis. The study applied logarithm to make alteration to the dataset.

The Iris dataset was prepared using these typical data preparation methods.

3.1.2 Data Mining Approaches in Machine Learning

The major data mining approaches used are:

i. Training data
This is one of the most important subsets, comprising over 60% of the entire dataset. This dataset is used to initially train the model. It tells the algorithm what to look for in the data.

ii. Validation data
This subset represents around 20% of the whole dataset to evaluate all of the model's parameters after the training phase. The validation data is actual data for identifying any shortcomings in the model to assess how well or poorly the model fits the data.

iii. Test data
At the very last stage of the training process, this subset which includes the final 20% of the dataset is introduced. The data in this group is not known to the model and is used to evaluate its precision. This dataset shows how much the model has learned from the prior two subsets. Fig. 2 and Fig. 3 show the detailed representation of the architectural model for image classification and process flow for image classification respectively for this study.

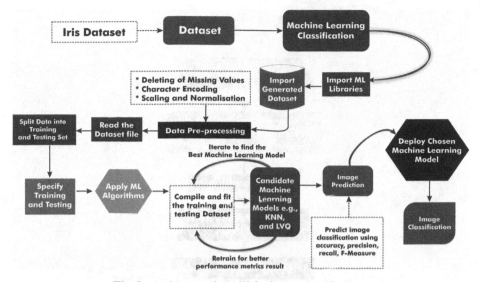

Fig. 2. Architectural model for image classification.

3.2 Mathematical Modelling

Various mathematical expressions used in KNN and LVQ to carry out successful operations are discussed in this section. Mathematical modelling are ways of expressing operations or activities using mathematical expressions.

3.2.1 Mathematical Expressions

There are two primary components to the KNN mathematical expressions.

1. Distance Calculation

The initial step in KNN is to determine how far apart each training example and the fresh input are from one another. Typically, a distance metric like Euclidean distance is used for this, which is denoted by the following definition:

$$d(x, y) = ((x_1 - y_1)2, (x_2 - y_2)2, \ldots \ldots, (x_n - y_n)2) \tag{11}$$

where: n = number of features, x = input vector and y = training vector

Once the distances have been determined, the next step is to locate the k nearest neighbors and use them to generate a forecast. This process is known as KNN classification. When using KNN classification, the prediction is made by majority vote, and the new input is given the class label that is most frequently used by its k closest neighbors. KNN classification's mathematical expression is represented as follows:

$$y = majority\_vote(y_1, y_2, \ldots \ldots, y_k) \tag{12}$$

where y_i is the *ith* nearest neighbor's class label.

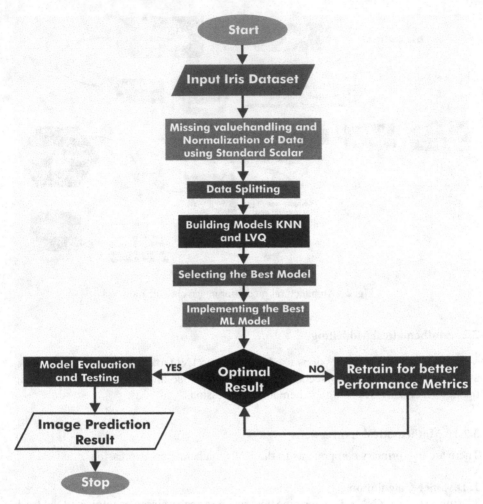

Fig. 3. Process flow for image classification.

In a manner similar to KNN regression, the prediction is made by averaging the k nearest neighbors' target values:

$$y = mean(y_1, y_2, \ldots\ldots, y_k) \tag{13}$$

These are the fundamental mathematical formulas for KNN, and any computer language can utilize them to build the method.

2. Euclidean Distance

The Euclidean distance, a measurement of the straight-line distance between two locations in a multidimensional space, is widely used in the k-nearest neighbors (KNN) technique [19]. In multidimensional space, Fig. 4 depicts the Euclidean distance between two points.

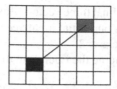

Fig. 4. Euclidean distance between two points in a multidimensional space.

The Euclidean distance between a test sample and each training sample is determined in the context of KNN, and the k training samples with the closest Euclidean distances to the test sample are denoted as the test sample's "nearest neighbors."

The test sample is then predicted by these nearest neighbors by majority vote or by averaging their corresponding goal values. Between two n-dimensional vectors, x and y, the Euclidean distance is defined by [20]:

$$d(x, y) = \sqrt{\sum\nolimits_{i=1}^{n} (x_i - y_i)^2} \tag{14}$$

where x_i and y_i, respectively, are the ith components of the x and y vectors.

This distance is the most popular one because the Python SkLearn package uses it as the default measure for K-Nearest Neighbor. It is a calculation of the actual straight-line distance between two points in Euclidean space Anil Gokte (2020).

3.2.2 Mathematical Expression of Learning Vector Quantization

Consider the target class for the following five input vectors in Table 2.

Table 2. Five input vectors and their target class.

Vector	Class label
[0 0 1 1]	1
[1 0 0 0]	2
[0 0 0 1]	3
[1 1 0 0]	4
[0 1 1 0]	5

There are two target classes (1 and 2) and four input components (x_1, x_2, x_3, x_4) in each input vector. Assigning weights based on the class is a good idea. The first two vectors, w1 = [0 0 1 1] and w2 = [1 0 0 0], can be utilized as weight vectors because there are two target classes.

$$W = \begin{bmatrix} 0 & 1 \\ 0 & 0 \\ 1 & 0 \\ 1 & 0 \end{bmatrix} \tag{15}$$

Training can be done with the final three vectors. Take 0.1 for the learning rate. Take a look at the third vector, which is our initial input vector.

Input vector: [0 0 0 1] Target class: 2

The subsequent step is to compute the Euclidean distance. The method is:

$$D(j) = \sum_{i=1}^{n} \left(w_{ij} - x_i\right)^2 \tag{16}$$

where, w_{ij} is the weight, x_i is the input vector component. Now, the distance of the input unit from the initial to the subsequent weight vectors, $D(1)$ and $D(2)$ respectively can be computed.

$$D(1) = (0 - 0)^2 + (0 - 0)^2 + (1 - 0)^2 + (1 - 1)^2 = 1 \tag{17}$$

$$D(2) = (1 - 0)^2 + (0 - 0)^2 + (0 - 0)^2 + (0 - 1)^2 = 2 \tag{18}$$

Here, $D(1)$ has a smaller value compare to $D(2)$ with leading index $J = 1$

Weight updating can be performed, because the target class has an unequal value as J. The following approaches are followed:

$$W_{11}(new) = w_{11}(old) - \alpha[x_1 - w_{11}(old)] = 0 - 0.1[0 - 0] = 0 \tag{19}$$

$$W_{21}(new) = w_{21}(old) - \alpha[x_2 - w_{21}(old)] = 0 - 0.1[0 - 0] = 0 \tag{20}$$

$$W_{31}(new) = w_{31}(old) - \alpha[x_3 - w_{31}(old)] = 1 - 0.1[0 - 1] = 1.1 \tag{22}$$

$$W_{41}(new) = w_{41}(old) - \alpha[x_4 - w_{41}(old)] = 1 - 0.1[1 - 1] = 1 \tag{22}$$

The updated weight vector will be:

$$W = \begin{bmatrix} 01 \\ 00 \\ 1.10 \\ 10 \end{bmatrix} \tag{23}$$

$$D(1) = (0 - 1)^2 + (0 - 1)^2 + (1.1 - 0)^2 + (1 - 0)^2 = 4.21 \tag{24}$$

$$D(2) = (1 - 1)^2 + (0 - 1)^2 + (0 - 0)^2 + (0 - 0)^2 = 1 \tag{25}$$

Here, the value of $D(2)$ is not up to the value of $D(1)$, and the leading index is $J = 2$.

The weight is also updated because the target class is not up to J. This can be achieved as follows:

$$W_{12}(new) = w_{12}(old) - \alpha[x_1 - w_{12}(old)] = 1 - 0.1[1 - 1] = 1 \tag{26}$$

$$W_{22}(new) = w_{22}(old) - \alpha[x_2 - w_{22}(old)] = 0 - 0.1[1 - 0] = -0.1 \tag{27}$$

$$W_{32}(new) = w_{32}(old) - \alpha[x_3 - w_{32}(old)] = 0 - 0.1[0 - 0] = 0 \qquad (28)$$

$$W_{42}(new) = w_{42}(old) - \alpha[x_4 - w_{42}(old)] = 0 - 0.1[0 - 0] = 0 \qquad (29)$$

The updated weight vector will be:

$$W = \begin{bmatrix} 01 \\ 0 - 0.1 \\ 1.10 \\ 10 \end{bmatrix} \qquad (30)$$

Input vector: [0 1 1 0] Target class: 1

$$D(1) = (0 - 0)^2 + (0 - 1)^2 + (1.1 - 1)^2 + (1 - 0)^2 = 2.01 \qquad (31)$$

$$D(2) = (1 - 0)^2 + (-0.1 - 1)^2 + (0 - 1)^2 + (0 - 0)^2 = 3.21 \qquad (32)$$

$D(1)$ is smaller compared to $D(2)$ with leading index $J = 1$. The weight is updated because the target class and J have the same value.

$$W_{13}(new) = w_{13}(old) + \alpha[x_1 - w_{13}(old)] = 0 + 0.1[0 - 0] = 0 \qquad (33)$$

$$W_{23}(new) = w_{23}(old) + \alpha[x_2 - w_{23}(old)] = 0 + 0.1[1 - 0] = 0.1 \qquad (34)$$

$$W_{33}(new) = w_{33}(old) + \alpha[x_3 - w_{33}(old)] = 1 + 0.1[1.1 - 1] = 1.09 \qquad (35)$$

$$W_{43}(new) = w_{43}(old) + \alpha[x_4 - w_{42}(old)] = 1 + 0.1[0 - 1] = 0.9 \qquad (36)$$

The updated weight vector will be:

$$W = \begin{bmatrix} 0 & 1 \\ 0.1 & -0.1 \\ 1.09 & 0 \\ 0.9 & 0 \end{bmatrix} \qquad (37)$$

Weights have been applied to all three input vectors. This symbolise the final stage for the first iteration. Until the winning vectors equate the target class, n number of epochs can be computed.

4 Results and Discussion

This section explains the performance metrics obtained from KNN and LVQ as well as a comparison of their outcomes on accuracy, precision, recall, and F1 Score.

4.1 Mathematical Expressions of the Performance Evaluation of the Models

Four performance evaluation metrics used in measuring KNN and LVQ performance which are accuracy, recall, F1 score as well as precision. These metrics are used to assess a k-NN classifier's performance and aid in figuring out the ratio of false positives to false negatives. A classifier that predicts accurately and has a low rate of false positives and false negatives would have high accuracy, precision, recall, and F1 scores. These performance evaluation measures were also used for LVQ algorithms.

i. Accuracy: is the correctly classified instances percentage [22]. The accuracy of a k-NN classifier is calculated by dividing the percentage of accurate predictions it generates by the total amount of guesses. The following illustrates how it can be mathematically expressed:

$$\text{Accuracy} = \frac{number\ of\ correct\ predictions}{total\ number\ of\ predictions} \tag{38}$$

ii. Precision: this is the classified modules number that are prone to fault and that are truly fault-prone modules [22]. It is determined by dividing the total number of correctly predicted positive outcomes by the actual positive outcomes. It is stated as follows:

$$\text{Precision} = \frac{True\ Positives}{(True\ Positves\ +\ False\ Positives)} \tag{39}$$

iii. Recall: The proportion of correct positive predictions to correct positive cases is known as recall. It is illustrative of:

$$\text{Recall} = \frac{True\ Positives}{(True\ Positves\ +\ False\ Negatives)} \tag{40}$$

iv. F1 Score: The F1 Score can be expressed as the harmonic mean of recall and precision.

$$\text{F1 Score} = \frac{2 * (Precision * Recall)}{Precision + Recall} \tag{41}$$

4.2 Comparison of KNN and LVQ Performance Evaluation

KNN performance was measured using the four (4) evaluation metrics with 96.67% accuracy, 1.00 precision, 0.89 recall, and 0.94 F1 Score. Also, LVQ performance was measured using the four (4) evaluation metrics where the accuracy is 80%, precision is 1.00, recall is 0.80 and F1 Score is 0.80. Regarding the comparative analysis, the precision result from KNN and LVQ is the same which is 1.00. The graphical illustrations are displayed in Fig. 5 while other metrics outcomes are represented in Table 3.

Table 3. Comparison of KNN and LVQ performance evaluation.

Algorithms	Accuracy (%)	Precision	Recall	F1 Score
K Nearest Neighbour	96.67	1.00	0.89	0.94
Learning Vector Quantization	80	1.00	0.80	0.80

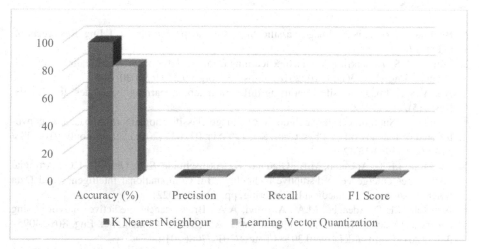

Fig. 5. Comparison of KNN and LVQ accuracy, precision, recall, and F1 score results.

4.3 Comparison of the Proposed Model with Other Studies

This section involves the comparison of this study with the results from other studies. Table 4 shows the comparison detailing the methodology, model used, and results.

Table 4. Comparison of KNN performance evaluation with other papers.

Author	Methodology	Model	Accuracy
[23]	Comparison of Iris dataset using Classification techniques	k-NN	93%
[24]	Classifying flowers images by using different classifiers in Orange	k-NN	82%
Proposed Model	**Comparison of Iris Dataset using k-NN and LVQ**	**k-NN**	**96.67%**

5 Conclusion

The study's comparison of LVQ and KNN models revealed that both algorithms while performing differently are useful for classification tasks. On the test set, the LVQ model had an accuracy of 80% whereas the KNN model had an accuracy of 96.67%. This shows that the KNN model can generalize new data more effectively. The algorithms were trained and tested using a 70:30 train-test split ratio. The study emphasizes how important it is to carefully choose an algorithm for certain classification tasks because every algorithm has strengths and shortcomings of its own. The study therefore, recommends for further studies the implementation of deep learning using larger dataset.

References

1. Bharadi, V., Rode, N.N.: Image classification using deep learning. Int. J. Eng. Res. Technol. **6**(11), 17–19 (2017)
2. Loussaief, S., Abdelkrim, A.: Machine learning framework for image classification. Adv. Sci. Technol. Eng. Syst. **3**(1), 1–10 (2018). https://doi.org/10.25046/aj030101
3. Alaba, S.Y.: Image classification using different machine learning techniques. Int. J. Adv. Res. **7**(5), 1–5 (2019)
4. Jiantao, Z., Shumin, C.: Research on flower image classification algorithm based on convolutional neural network. J. Phys. Conf. Ser. **1994**(1), 012034 (2021). https://doi.org/10.1088/1742-6596/1994/1/012034
5. Akinsola, J.E.T., Adeagbo, M.A., Oladapo, K., Akinsehinde, S.A., Onipede, F.O.: Artificial intelligence emergence in disruptive technology. In: Computational Intelligence and Data Science Paradigms Biomedical Engineering, pp. 63–99 (2022)
6. Akinsola, J.E.T., Adeagbo, M.A., Awoseyi, A.A.: Breast cancer predictive analytics using supervised machine learning techniques. Int. J. Adv. Trends Comput. Sci. Eng. **8**(6), 3095–3104 (2019). https://doi.org/10.30534/ijatcse/2019/70862019
7. Awoseyi, A.A., Akinsola, J.E.T., Oladoja, M.O., Adeagbo, M.A., Adebowale, O.O.: Hybridization of decision tree algorithm using sequencing predictive model for COVID-19. In: Mondal, M.R.H., Kose, U., Prasath, V.B.S., Podder, P., Bharati, S., Kamruzzaman, J. (eds.) Emerging Technologies for Combating Pandemic: AI, IoMT and Analytics. CRC Press, Taylor & Francis Group, USA (2021)
8. Akinsola, J.E.T., Awodele, O., Idowu, S.A., Kuyoro, S.O.: SQL injection attacks predictive analytics using supervised machine learning techniques. Int. J. Comput. Appl. Technol. Res. **9**(4), 139–149 (2020). https://doi.org/10.7753/ijcatr0904.1004
9. IBM, What is the k-nearest neighbors algorithm? IBM (2022)
10. Iparraguirre-Villanueva, O., Espinola-Linares, K., Flores Castañeda, R., Cabanillas-Carbonell, M.: Application of machine learning models for early detection and accurate classification of Type 2 Diabetes. Diagnostics (Basel, Switzerland) **13**(14), 2383 (2023). https://doi.org/10.3390/diagnostics13142383
11. Anowar, F., Sadaoui, S., Selim, B.: Conceptual and empirical comparison of dimensionality reduction algorithms (PCA, KPCA, LDA, MDS, SVD, LLE, ISOMAP, LE, ICA, t-SNE). Comput. Sci. Rev. **40**, 1–13 (2021). https://doi.org/10.1016/j.cosrev.2021.100378
12. LaViale, T.: Deep Dive on KNN_ Understanding and Implementing the K-Nearest Neighbors Algorithm. Arize AI (2023)
13. Baliyan, M.: Learning Vector Quantization, GeeksforGeeks (2023)
14. Biehl, M., Ghosh, A., Hammer, B.: Learning vector quantization: the dynamics of winner-takes-all algorithms. Neurocomputing **69**, 660–670 (2006). https://doi.org/10.1016/j.neucom.2005.12.007
15. Turing, How to Implement Learning Vector Quantization from Scratch with Python, Turing (2023)
16. Engelsberger, A., Thomas, V.: Quantum computing approaches for vector quantization—current perspectives and developments. In: MDPI (2023)

17. Nasir, V., Schimleck, L., Abdoli, F., Rashidi, M., Sassani, F., Avramidis, S.: Quality control of thermally modified Western Hemlock Wood using near-infrared spectroscopy and explainable machine learning. In: MDPI (2023)
18. Shemir, J.: Quick Guide to Datasets for Machine Learning in 2023 (2023)
19. Yasmeen, R.: K-Nearest Neighbor (KNN) Algorithm in Machine Learning, Medium (2021)
20. Byjus, Euclidean Distance - Definition, Formula, Derivation & Examples, Byjus (2023)
21. Gokte, S.A.: Most Popular Distance Metrics Used in KNN and When to Use Them, Praxis Business School (2020)
22. Akinsola, J.E.T., Awodele, O., Kuyoro, S.O., Kasali, F.A.: Performance evaluation of supervised machine learning algorithms using multi-criteria decision making techniques. In: International Conference on Information Technology in Education and Development (ITED), pp. 17–34 (2019)
23. Prathima, P., Ranjith, K.T.: Comparison on iris dataset using classification techniques. J. Emerg. Technol. Innnovative Res. **8**(8), 315–319 (2021)
24. Sajwan, V., Ranjan, R.: Classifying flowers images by using different classifiers in orange. Int. J. Eng. Adv. Technol. **8958**(6), 1057–1061 (2019). https://doi.org/10.35940/ijeat.F1334.0986S319

Enhancing IDC Histopathology Image Classification: A Comparative Study of Fine-Tuned and Pre-trained Models

Anusree Kanadath[✉], J. Angel Arul Jothi, and Siddhaling Urolagin

Department of Computer Science, Birla Institute of Technology and Science Pilani, Dubai Campus, Dubai International Academic City, 345055 Dubai, UAE
{p20180904,angeljothi,siddhaling}@dubai.bits-pilani.ac.in

Abstract. Invasive ductal carcinoma (IDC) is a type of breast cancer that affects adult women all around the world. This cancer starts in the duct cells of the breast, spreads through the lymph system, and eventually affects nearby organs and bones. It's crucial for physicians to correctly identify the various forms of breast cancer. Instead of doing this manually, it's better to use computer programs because it saves time and reduces mistakes. This study presents a computer-assisted diagnosis method that utilizes deep convolutional neural networks to classify IDC histopathology images. These networks are trained using two forms of transfer learning: feature extraction and fine-tuning. In this study IDC classification is done using 4 well known deep learning networks, Xception, DenseNet169, ResNet101 and MobileNetV2. The dataset used is a publicly available IDC dataset containing 168 whole slide images. The evaluation results show that the fine-tuned models give better classification results than feature extractor models for IDC histopathology image classification.

Keywords: Deep learning · Transfer learning · Histopathology · Image classification · Fine-tuned models · Computer aided diagnosis

1 Introduction

Invasive ductal carcinoma (IDC) is the most prevalent form of breast cancer, accounting for approximately 80% of all breast cancer cases in women. The term 'invasive' signifies that the cancer has extended into the adjacent breast tissues. 'Ductal' implies that the cancer originated within the milk ducts, the channels responsible for transporting milk from the lobules to the nipple. Lastly, 'Carcinoma' denotes any cancer that initiates in the skin or other tissues covering internal organs, including breast tissues. IDC is characterized by the transformation of abnormal cells within the milk duct lining, leading to their invasion of breast tissue beyond the confines of the duct walls. Once that happens, the cancer cells can spread. In the United States, it is anticipated that 297,790 cases of invasive breast cancer and 55,720 cases of non-invasive breast cancer will be diagnosed in the year 2023 [2]. In the last few decades, the number of women

K. K. Patel et al. (Eds.): icSoftComp 2023, CCIS 2031, pp. 164–176, 2024.
https://doi.org/10.1007/978-3-031-53728-8_13

diagnosed with invasive breast cancer has increased by approximately 0.5% per year. For the year 2023, estimates suggest that 2800 men in the United States will also be diagnosed with invasive breast cancer. Tragically, it is anticipated that breast cancer will result in approximately 43700 deaths in the United States. Among these, the vast majority, around 43170, are expected to affect women, with a smaller number, approximately 530 affecting men.

Timely identification of invasive carcinoma plays a pivotal role in cancer treatment. Detecting invasive carcinoma at an early stage substantially enhances the prospects of successful treatment and long-term survival. As invasive carcinomas progress, it is typically more difficult to treat, necessitating more aggressive therapies and posing greater health risks for the patient. Not only does early detection increases the likelihood of complete tumor removal, but it also allows for a wider range of treatment options that may be less invasive and have fewer adverse effects.

Researchers have made significant advances in utilizing deep learning (DL) algorithms to aid in cancer screening, including breast cancer. These DL algorithms aid pathologists in rapidly analyzing histopathology images and diagnosing cancer. Despite the remarkable success of DL algorithms, their seamless integration into digital pathology faces significant obstacles. Some of these challenges include the absence of the necessary labeled data for complex deep learning models, the texture variation of the tissue types and the vast dimensionality of whole slide images (WSIs) with common image sizes exceeding 50000×50000 pixels.

Transfer learning is one of the effective solutions for histopathology image processing [4,14,15]. Transfer learning is the process of applying a model that has been trained on a large dataset for a specific task to a similar task, despite having a smaller dataset. It offers numerous benefits, including the reduction of training time, the improvement of output accuracy, and the requirement for less training data. Negative transfer and overfitting are the two major disadvantages of transfer learning.

This paper focuses on transfer learning-based approaches for classifying histopathology image regions as IDC+ve or IDC-ve. The pretrained deep models are reused for feature extraction approach and fine-tuned approach. 4 deep learning models, XceptionNet, DenseNet169, ResNet101 and MobileNetV2 pretrained on the ImageNet dataset were used. An IDC dataset with 168 WSIs that is publicly available was used for this study. The detailed comparative analysis helps the researchers to identify the best transfer learning models for IDC histopathology image classification.

The remainder of this paper is structured as follows: Sect. 2 provides an overview of the relevant literature. Section 3 describes the methodology used in this study. Section 4 describes in detail the experimental setup. The results and analysis are presented in Sect. 5. Section 6 is the conclusion of the paper.

2   Literature Survey

In this section, we review several recent papers related to the automated detection of IDC in breast cancer using deep learning techniques. These papers demonstrate the advancements in this field and the ongoing efforts to enhance IDC classification accuracy and efficiency.

Andrew janowczyk and Anant madabhushi [12] have developed and implemented a deep learning model for a variety of digital pathology tasks, including segmentation, detection, and classification. The study achieved an F1 score of 0.7648 in IDC detection task. Aiza and Alexander [18] presented an improved convolutional neural network (CNN) architecture for predicting IDC. Their model demonstrated remarkable performance with an F1 score of 85.28% and a balanced accuracy of 85.41%, surpassing previous deep learning approaches. This paper emphasized the significance of CNN enhancements for accurate IDC detection.

Jianfei zhang et al. [22] proposed a method that merge a multi-scale residual CNN (MSRCNN) and support vector machine (SVM) for IDC detection. The approach demonstrated an average accuracy of 87.45%, average balanced accuracy of 85.7%, and an average F1 score of 79.89% after 5-fold cross-validation. Avishek and Sunanda [7] implemented a CNN model for breast cancer classification. The model demonstrated a classification accuracy of 78.4% when evaluated on the IDC dataset. Mohammad et al. [3] investigated two approaches for IDC classification: a baseline CNN model and transfer learning using the VGG16 CNN model. The baseline model achieved an F1 score of 83% and an accuracy of 85%. Notably, transfer learning through feature extraction produced superior classification results compared to the baseline model.

Justin et al. [20] investigated a range of CNN architectures for automated breast cancer detection. They assessed four different architectures using a substantial dataset and achieved remarkable results with one particular fine-tuned CNN architecture. This model yielded impressive results, including an F1 score of 92%, a balanced accuracy of 87%, and an accuracy of 89%. The study identified a finely tuned CNN architecture that consistently delivered outstanding performance. Érika et al. [5] emphasized the advantages of using deep learning for IDC detection. Their 3-hidden-layer CNN, with data balancing, achieved both accuracy and an F1-Score of 0.85.

In conclusion, the literature survey has provided a comprehensive overview of the state of research in IDC image classification using deep learning. While significant advancements have been achieved in attaining elevated accuracy and F1-scores, there remains an ongoing need for deeper exploration into the most efficient approaches like fine-tuning and feature extractor models. This encompasses a comprehensive examination of various model architectures and transfer learning techniques to enhance their efficacy.

3   Materials and Methods

3.1   Transfer Learning

Machine learning and deep learning have revolutionized computer vision, natural language processing, and speech recognition by achieving success at complex
tasks. However, these models often need huge, high-quality datasets and significant computational resources. The performance of DL models is critically
dependent on the availability of an adequate volume of accurately labeled training data [17]. While there is an abundance of labelled data for natural images, a
lack of annotated medical images presents a significant challenge in the domain of
medical image analysis. This lack of training data has the potential to hinder the
effectiveness of deep learning models. Therefore, transfer learning has emerged as
a viable alternative to conventional DL approaches, providing a valuable solution
to enhance model performance and overcome data limitations [9].

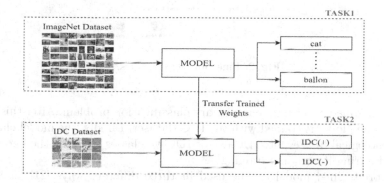

Fig. 1. General diagram of transfer learning for image classification.

Pre-trained models are DL models that have been used to solve one problem
using a large dataset and then reused to solve another similar problem with a
smaller dataset. Transfer learning is the process of transferring a pre-trained
model's weights to solve another problem. Transfer learning saves training time,
improves neural network performance, and reduces the demand for data. These
advantages collectively contribute to the widespread popularity of transfer learning as a powerful machine learning method. The general diagram of transfer
learning is given in the Fig. 1. The deep learning model is initially trained using
the ImageNet visual database, which contains more than 14 million images, with
the objective to classify images into 1000 distinct classifications. After training,
the model's weights are set to their optimal values, resulting in a model that has
been effectively learned. This pre-trained model is then used within a context
for transfer learning. The pre-trained model is repurposed for binary classification in order to solve the IDC histopathology image classification problem. The

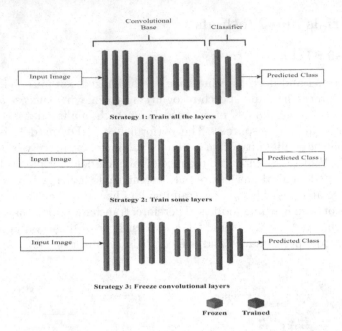

Fig. 2. Fine-tuning strategies.

model is modified according to the binary classification problem. After this modification, the model is trained with the IDC dataset. Finally, the model classifies histopathological images into IDC+ve or IDC-ve classes by using the knowledge it gained during its original training.

The pre-trained model can be used in three different ways [21], as shown in Fig. 2. The pre-trained model consists of a pre-trained convolutional base followed by a classifier. The convolutional base is comprised of a series of convolutional and pooling layers designed to extract image features. In contrast, the classifier is comprised of fully connected layers whose primary function is to classify images based on the extracted features [13]. The first approach is to use the IDC dataset to train the whole model. In this method, the architecture of the model that has already been trained is preserved, while training is tailored to suit the specific requirements of the IDC dataset. Nonetheless, this method requires a large dataset and substantial computational resources. The second approach is to freeze some layers of the convolutional base while training others. The lower layers capture general characteristics, whereas the upper layers focus on problem-specific characteristics. During the training process, certain layers can be "frozen" and kept unchanged by altering their layer weights. This method is especially useful when working with limited datasets or models with numerous parameters. This method helps the model to acquire both general and task-specific features, which may result in improved performance. This adaptability is a significant advantage, but it requires more computational resources. The third approach involves freezing all the layers of the convolutional base, in its original form. This strategy is useful when computational resources are scarce, dataset size is small, or the pre-trained model has already demonstrated proficiency in solving a problem closely related to the target task. This approach can be computationally efficient, as it avoids the need to train all layers from scratch, making it more feasible in resource-limited situations.

3.2 Pre-trained Models

Several image classification models are training on extensive image datasets, including the widely known ImageNet. Some of the most well-known pre-trained classification models are AlexNet, VGG, GoogLeNet, ResNet, DenseNet, MobileNet, EfficientNet, Xception, NASNet, SqueezeNet, ShuffleNet, etc. [19]. 4 popular pre-trained models that have shown promising results for medical image classification such as DenseNet, ResNet, MobileNet, and XceptionNet are used in this work. This next section provides an overview of the characteristics of these pre-trained models.

ResNet: ResNet, commonly referred to as residual networks, represents a notable breakthrough in the domains of deep learning and computer vision [8]. As networks grow deeper with more layers, they often encounter the vanishing gradient problem, affecting effective training [16]. ResNet proposed a solution with the implementation of residual blocks, which are alternatively referred to as skip connections or shortcut connections. These connections provide alternative pathways for data and gradients to flow thus making training possible.

Figure 3 depicts the fundamental building block of the resnet, known as the 'residual block.' A residual block consists of two convolutional layers (Conv) accompanied by batch normalization (BN) and ReLU activation functions. The input feature map is denoted as X, while $F(X)$ represents the output obtained after passing through the two convolutional layers followed by BN and ReLU layers. Then the final output $H(X)$ from the residual block is defined by Eq. 1.

$$H(X) = F(X) + X \tag{1}$$

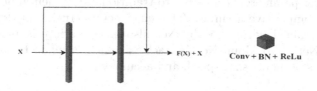

Fig. 3. Architecture of a residual block.

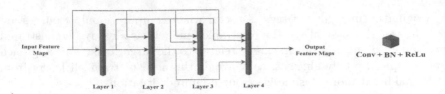

Fig. 4. Architecture of a DenseNet Block.

DenseNet: The DenseNet model developed by Huang et al. shown remarkable classification performance in 2017 when applied to publically available image datasets such as CIFAR-10 and ImageNet [11]. In the DenseNet architecture, every layer is connected to the successive layers within the network. This means that the features acquired by any layer are readily shared throughout the entire network, creating an enhanced information flow. Consequently, this architecture significantly improves the efficiency of training deep networks, all the while enhancing model performance. Furthermore, the presence of dense connections plays a role in mitigating overfitting, particularly on tasks involving smaller datasets. Figure 4 depicts a fundamental building block in the DenseNet architecture, referred to as a DenseNet block. In this block, the output of each convolutional layer is not only passed forward to the next immediate layer but also serves as input to every subsequent convolutional layer within the same block.

MobileNet: MobileNet is a family of lightweight deep neural network architectures designed for fast and effective deployment on mobile and embedded devices [10]. MobileNet's efficiency is based on the concept of depthwise separable convolution, which divides the standard convolution operation into two distinct steps: depthwise convolution and pointwise convolution. In depthwise convolution, a single filter is applied per input channel, resulting in a substantial reduction in computational load. The subsequent pointwise convolution combines the outcomes of the depthwise convolution to produce feature maps. This technique significantly reduces the computational overhead while preserving model accuracy. MobileNet architectures are renowned for their parameter efficiency. The use of depthwise separable convolution and reduced model size means they have significantly fewer parameters compared to traditional deep neural networks while maintaining competitive accuracy. This efficiency is crucial for deployment on resource-constrained devices. MobileNet also has a variety of model versions, from MobileNetV1 to MobileNetV3, each of which is made to meet different needs in terms of model size, speed, and accuracy.

XceptionNet: XceptionNet, often known as "Extreme Inception," represents a significant breakthrough in the field of deep learning and computer vision [6]. The basic concept behind XceptionNet is depthwise separable convolutions, similar to the approach used in MobileNet. The application of depthwise separable convolutions reduces model parameters and improves computational efficiency. The design of XceptionNet was inspired by the multi-scale feature extraction capabilities of the Inception architecture. However, Xception takes this concept to an extreme by implementing depthwise separable convolutions across all Inception modules. This deep and efficient architecture enables XceptionNet to capture intricate patterns and features in data while achieving impressive computational efficiency.

3.3   Proposed Model

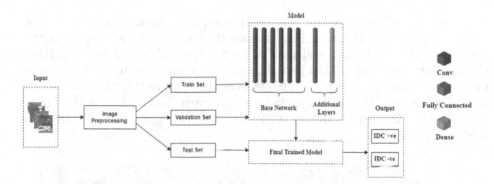

Fig. 5. An illustration of Proposed model for IDC image classification

Figure 5 provides a detailed overview of the IDC image classification process. The process begins by acquiring images from the IDC dataset, which are then subjected to image preprocessing. Each image is resized to 48×48 dimensions during this preprocessing phase. To address class imbalances in the dataset, oversampling techniques such as SMOTE are applied to the images. After achieving class balancing, the dataset is divided into three sets: the training set, the test set, and the validation set. The classification model is trained using images from the training and validation sets as well as their respective class labels. These classification models are pre-trained models that consist of a base network followed by additional layers tailored for the specific classification task. After the training phase, the trained model is prepared for testing. During the testing phase, images from the test set are fed into the trained model, which predicts whether each image is IDC+ve or IDC-ve.

The top layers of each pre-trained model is removed treating the remaining architecture as the base network for the proposed model. The specific number of top layers removed was determined through experimentation and analysis of the model's architectures. After that, a series of batch normalization, dropout, and fully connected layers are added to this network. These additions were thoroughly selected to optimize the performance of the model. In our experiments on the IDC dataset, we investigate two distinct approaches. The first approach is the feature extraction approach, which involves freezing all layers of the base network while training the remaining layers using a balanced IDC dataset. It is decided to freeze all layers so that the general feature representations that the pre-trained model had learned would remain the same. In contrast, the second approach is the fine-tuning approach, which entails freezing specific layers of the base network, enabling the remaining layers to undergo training using the balanced IDC dataset. It also enables the capture of IDC specific features while retaining general knowledge from the pre-trained layers.

4   Experimental Setup

The IDC dataset employed in this study comprises digitized histopathology slides collected from 162 individuals diagnosed with IDC at the Hospital of the University of Pennsylvania and the Cancer Institute of New Jersey [1]. From this dataset, a total of 277,524 patches, each measuring 50×50×3 (RGB), were extracted. Among these patches, 198,738 were identified as IDC-ve, while 78,786 were classified as IDC+ve. Sample images from each class is given in the Fig. 6. First 6 images belongs to IDC -ve classes and last 6 images belongs to IDC +ve classes.

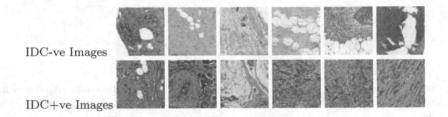

IDC-ve Images

IDC+ve Images

Fig. 6. Sample images from IDC dataset.

During the training phase, the IDC dataset was split 80:10:10 into training, testing, and validation sets. The model was trained for 100 epochs using the Adam optimizer. To evaluate the performance of the model, the binary cross-entropy loss function was used, and a batch size of 128 was chosen to process the 48×48 histopathology images efficiently. Accuracy, precision, recall, and F1 score were employed as evaluation metrics. The experiments were conducted on a computer with an Intel(R) Xeon(R) W-2123 CPU, 16 GB of RAM, and a 1 TB hard drive. The TensorFlow framework was employed for implementing the model, and the code was developed using the Python programming language.

5   Results and Discussion

In this section, we present the experimental results and analyze the outcomes of two distinct methods applied to the IDC dataset: the feature extraction approach and the fine-tuning approach. We selected four deep classification models for our experiments and evaluated their performance on a balanced IDC dataset. A comparative analysis was conducted by considering both the number of parameters to be trained and the classification accuracy.

5.1   Parameter Analysis of Deep Learning Models

Table 1 provides a comprehensive overview of the trainable and non-trainable parameters associated with the deep models used in our experiments. The feature extractor models were directly applied to the dataset without any structural modifications, with the only alteration being the addition of fully connected layers at the end. Consequently, feature extractor models tend to have a high count of non-trainable parameters while keeping the number of trainable parameters relatively low. Among the various feature extractor models, MobileNet stands out with an exceptionally low number of parameters. This is attributed to its design, which prioritizes efficiency for lightweight devices by employing depthwise separable convolution operations to reduce parameter count. In contrast, ResNet and its variants tend to have a notably higher number of parameters due to their utilization of residual networks within their architecture.

Similarly fine-tuning involves taking an existing deep model and adjusting it to better suit a specific task or dataset. Unlike feature extractor models, fine-tuning allows for more flexibility in modifying the model's architecture, including unfreezing and retraining certain layers. By keeping most of the model's parameters frozen, fine-tuning requires training only a fraction of the total parameters, making the process more efficient and faster compared to training an entirely new model. Among the various fine-tuned models, ResNet101 has a high number of parameters to train. When comparing fine-tuning with feature extraction approach, one notable difference is that fine-tuning typically involves training more parameters than feature extraction approach. This allows the model to adapt its representations and features for a specific task, making it more suitable for the target application.

Table 1. Details of trainable and non-trainable parameters (in million).

Models	Feature extraction		Fine-tuning	
	Non-trainable	Trainable	Non-trainable	Trainable
Xception	20.89	0.0369	14.07	6.82
DenseNet169	12.64	0.0033	11.66	0.9788
ResNet101	42.67	0.0164	28.20	14.46
MobileNetV2	2.26	0.0102	0.0387	2.22

5.2   Classification Results and Analysis

The classification results of the developed models are detailed in Table 2 and Table 3. Table 2 showcases the classification results of feature extractor models when applied to the IDC dataset, whereas Table 3 illustrates the classification outcomes of fine-tuned models on the same IDC dataset

Among the feature extractor models, Xception and DenseNet169 demonstrate the highest levels of efficiency. The Xception model achieved a training accuracy of 0.85 and a testing accuracy of 0.80, demonstrating a strong ability to match the training data and effectively adapt to new, unseen data. The model achieved 0.81 precision and 0.81 recall, resulting in an F1 score of 0.81. DenseNet169 exhibits excellent results, with a training accuracy of 0.84 and a testing accuracy of 0.83, respectively. Both its precision and recall are high at 0.84, yielding an F1 score of 0.84. On the other hand, ResNet101 and MobileNetV2 exhibit slightly lower levels of performance. The ResNet101 model achieved a training accuracy of 0.74 and a testing accuracy of 0.74. The precision and recall values of the model are both 0.75, leading to an F1 score of 0.74. The MobileNetV2 model achieved a training accuracy of 0.79 and a testing accuracy of 0.76. Additionally, the precision, recall and F1 score are 0.77.

Table 2. Classification results of feature extractor models on balanced dataset.

Models	Train_acc	Test_acc	Precision	Recall	F1
Xception	0.85	0.80	0.81	0.81	0.81
DenseNet169	0.84	0.83	0.84	0.84	0.84
ResNet101	0.74	0.74	0.75	0.75	0.74
MobileNetV2	0.79	0.76	0.77	0.77	0.77

Among the fine-tuned models, Xception and DenseNet169 exhibit the highest levels of performance across all metrics. DenseNet169 achieved a training accuracy of 0.99 and a testing accuracy of 0.91. It maintains a remarkable precision of 0.91 and a high recall of 0.90, resulting in an F1 score of 0.90, indicating its effectiveness in both accuracy and the balance between precision and recall. Xception achieved a training accuracy of 0.99, indicating an excellent fit to the training data, while maintaining a testing accuracy of 0.87. Its precision, recall, and F1 score are 0.88, showcasing a well-balanced performance. In contrast, ResNet101 and MobileNetV2, while still achieving high training accuracy, demonstrate slightly lower testing accuracies of 0.81 and 0.78, respectively. Their precision and recall values are also lower than those of Xception and DenseNet169. ResNet101 has a precision, recall, and F1 score of 0.83, while MobileNetV2 has a precision of 0.81, a recall of 0.79, and an F1 score of 0.79.

The comparison between fine-tuned models and feature extractor models on the IDC dataset clearly states the superiority of fine-tuned models in terms of the

Table 3. Classification results of fine-tuned models on balanced dataset.

Models	Train_acc	Test_acc	Precision	Recall	F1
Xception	0.99	0.87	0.88	0.88	0.88
DenseNet169	0.99	0.91	0.91	0.90	0.90
ResNet101	0.99	0.81	0.83	0.83	0.83
MobileNetV2	0.99	0.78	0.81	0.79	0.79

accuracy of classification. Fine-tuning feature extractor models allows architecture changes, especially in the top layers that provide task-specific predictions. This modification substantially improves classification accuracy and other performance metrics. Fine-tuned models integrate the large amount of information acquired from a feature extractor model with the domain-specific knowledge obtained from the target dataset. This combination increases the model's ability to generalize to new, unknown data and decreases the possibility of overfitting. Finely-tuned models have a greater number of parameters to train than feature extractor models, resulting in a modest increase in training time. As part of future work, the application of additional deep learning models to the IDC classification is planned. Furthermore, the models will be extended to address multiclass classification challenges within other histopathology datasets.

6   Conclusion

In this study, we conducted an automated IDC image classification task by utilizing transfer learning techniques. Four deep learning models, Xception, ResNet, MobileNet, and DenseNet, were employed in the IDC classification. Among these models, the fine-tuned DenseNet169 outperformed, achieving a test accuracy of 91% and an F-score of 90%. This research highlights the significance of fine-tuning deep learning models to improve the accuracy of IDC diagnosis. The future scope of this research includes a broader exploration of deep learning models in IDC classification and a focus on multiclass classification challenges in various histopathology datasets.

References

1. Invasive ductal carcinoma (IDC) histology image dataset. http://www.andrewjanowczyk.com/use-case-6-invasive-ductal-carcinoma-idc-segmentation/
2. Breast cancer facts and statistics (2023). https://www.breastcancer.org/facts-statistics
3. Abdolahi, M., Salehi, M., Showkatian, E., Reiazi, R.: Artificial intelligence in automatic classification of invasive ductal carcinoma breast cancer in digital pathology images. Med. J. Islamic Repub. Iran **34**, 140 (2020)
4. Ahmed, S., et al.: Transfer learning approach for classification of histopathology whole slide images. Sensors (Basel, Switzerland) **21**, 5361 (2021)

5. de Assis, É. G., do Patrocinio, Z.K., Nobre, C.N.: The use of convolutional neural networks in the prediction of invasive ductal carcinoma in histological images of breast cancer. Stud. Health Technol. Inform. **290**, 587-591 (2022)
6. Chollet, F.: XCeption: deep learning with depthwise separable convolutions. In: 2017 IEEE Conference on Computer Vision and Pattern Recognition (CVPR) (2017)
7. Choudhury, A., Perumalla, S.: Detecting breast cancer using artificial intelligence: convolutional neural network. Technol. Health Care **29**, 33–43 (2020)
8. He, K., Zhang, X., Ren, S., Sun, J.: Deep residual learning for image recognition. In: Proceedings of 2016 IEEE Conference on Computer Vision and Pattern Recognition, CVPR '16. IEEE (2016)
9. Hee, K., Cosa, A., Santhanam, N., Jannesari, M., Maros, M., Ganslandt, T.: Transfer learning for medical image classification: a literature review. BMC Med. Imaging **22**, 69 (2022)
10. Howard, A.G., et al.: MobileNets: efficient convolutional neural networks for mobile vision applications. CoRR (2017)
11. Huang, G., Liu, Z., van der Maaten, L., Weinberger, K.Q.: Densely connected convolutional networks. In: 2017 IEEE Conference on Computer Vision and Pattern Recognition (2017)
12. Janowczyk, A., Madabhushi, A.: Deep learning for digital pathology image analysis: a comprehensive tutorial with selected use cases. J. Pathol. Inf. **7**, 29 (2016)
13. Kandel, I., Castelli, M.: How deeply to fine-tune a convolutional neural network: a case study using a histopathology dataset. Appl. Sci. **10**, 3359 (2020)
14. Ikromjanov, K., Bhattacharjee, S., Hwang, Y.B., Kim, H.C., Choi, H.K.: Multiclass classification of histopathology images using fine-tuning techniques of transfer learning. J. Korea Multimedia Soc. **24**, 849–859 (2021)
15. Mormont, R., Geurts, P., Marée, R.: Comparison of deep transfer learning strategies for digital pathology. In: 2018 IEEE/CVF CVPRW (2018)
16. Pascanu, R., Mikolov, T., Bengio, Y.: On the difficulty of training recurrent neural networks. In: ICML'13, JMLR.org (2013)
17. Rashmi, R., Prasad, K., Udupa, C.: Breast histopathological image analysis using image processing techniques for diagnostic purposes: a methodological review. J. Med. Syst. **46**, 1–24 (2021)
18. Romano, A.M., Hernandez, A.A.: Enhanced deep learning approach for predicting invasive ductal carcinoma from histopathology images. In: 2019 2nd International Conference on Artificial Intelligence and Big Data (2019)
19. Wang, J., Zhu, H., Wang, S., Zhang, Y.: A review of deep learning on medical image analysis. Mob. Netw. Appl. **26**, 351–380 (2021)
20. Wang, J.L., Ibrahim, A.K., Zhuang, H., Muhamed Ali, A., Li, A.Y., Wu, A.: A study on automatic detection of idc breast cancer with convolutional neural networks. In: 2018 International Conference on Computational Science and Computational Intelligence (CSCI) (2018)
21. Yamashita, R., Nishio, M., Do, R.K.G., Togashi, K.: Convolutional neural networks: an overview and application in radiology. Insights Imaging **9**, 611–629 (2018)
22. Zhang, J., Guo, X., Wang, B., Cui, W.: Automatic detection of invasive ductal carcinoma based on the fusion of multi-scale residual convolutional neural network and SVM. IEEE Access **9**, 40308–40317 (2021)

Optimized Path Planning Techniques for Navigational Control of Mobile Robot Using Grass Fire Algorithm in Obstacle Environment

Vengatesan Arumugam(✉) ⓘ and Vasudevan Algumalai

Department of Mechanical Engineering, Saveetha School of Engineering, SIMATS,
Chennai 602105, India
venkatesana9006.sse@saveetha.com

Abstract. This article discusses the better path planning, and achieve the goal position point within the minimum distance reached for using from mobile robots. The identification of the shortest distance was carried out through the optimization techniques, using from the grassfire algorithm. The grassfire algorithm, which was based on the provided structure of different square boxes, was employed. The grid structure adopted was (6 columns x 7 rows), and 27% of the obstacles were fixed within the grid structure of the graph. The pseudo-code of the grassfire algorithm was implemented for the analysis of simulation results. This implementation was utilized to the shortest path between points. Different pathways were explored to reach point 10 from point 0, with the objective of determining the shortest distance. For the remaining grid structures of the square boxes, the distance was treated as infinity, and the distance was updated using the formula: n distance = current distance + 1. The V-REP simulation was utilized in the experimentation, employing the Khepera-III robot to navigate through various obstacles in the environment. The results of the experimental and simulation analyses demonstrated a 4.6% deviation in the start and goal position points.

Keywords: Mobile robot · V-Rep simulation · Grassfire algorithm · Obstacle environmental · Optimized path planning · Khepera-III

1 Introduction

In various fields, including the military, space exploration, emergency situations involving fire threats, medicine, and other fields, mobile robots are now frequently used. The robot carried out the tough tasks listed above effectively and without assistance from humans. The term "path planning" was created to address such a situation. Whether or not the robot is familiar with its surroundings, path planning necessitates the robot traveling a particular path. A mobile robot must safely navigate around the various barriers and obstructions it comes across while navigating, avoid hitting them, and choose the best route from one spot to another [1]. These parameters must be met in order for the transport robot to complete its primary mission of delivering goods to the desired location [2]. "Off-line, or global, and on-line, or local, path planning" issues for robots are the

K. K. Patel et al. (Eds.): icSoftComp 2023, CCIS 2031, pp. 177–189, 2024.
https://doi.org/10.1007/978-3-031-53728-8_14

two main categories [3]. Off-line planning takes place in a robot's familiar environment where immovable impediments are also present. In this kind of path planning, the algorithm must prepare the entire path with coordinates using a variety of approaches before the robot begins to move. While online path development happens as an impediment between the source and destination points, local path planning completely fails in an unfamiliar area. In both of these systems, the path is chosen based on the environmental sensor data [4]. It has a changing environment and moving obstacles. Since the mobile robot in this setting only has incomplete or unknown knowledge of its surroundings, it must first perceive its surroundings before moving. Many authors have examined the topic of path planning for mobile robots in depth, and they look at a number of options [5].

Safety, precision, and speed are the three key issues that need to be resolved in robot navigation (RN). Finding a collision-free path and adhering to the targeted path precisely are the safety and accuracy issues. Efficiency refers to the algorithm's capacity to repeatedly stop and turn robots. Time and energy have been wasted on this. Localization, path planning, cognitive mapping, and motion control are only a few of the various RN issue areas. One could that path planning is the most crucial problem. The goal of path planning is to determine the best, most direct, and collision-free path through a given environment from a starting point to an objective. A robot typically has multiple ways to accomplish a task, but the best [6]. Mobile robots' ability to navigate successfully in various applications mostly depends on their intelligence [7]. These surveys, however, don't go far enough to offer a thorough examination of each navigational method. This proposed survey study on mobile robot navigation seeks to identify the areas for future research and the potential for innovation in a specific field [8]. A group of requirements that must be completed in order to produce the path planning solution that will lead to the achievement of the goal [9]. Many approaches, such as the grassfire algorithm, have been put forth to address the path planning issue [10]. Grassfire methods are now often used in the field of path planning and have been enhanced based on application scenario requirements [11]. Some innovative intelligence optimisation algorithms, such the genetic algorithm, have excelled at handling path planning issues in recent years [12]. It was used in a variety of optimisation strategies, including the shortest point reached [13].To estimate the optimal pathway, they suggested optimising the grassfire method [14]. The grassfire algorithm is an approach that uses heuristics to determine a route from a given source to a given destination [15]. Robot path planning has been implemented using a combination of numerical iteration and pseudo code grassfire algorithms [16]. According to the experimental findings, the robot successfully reached its intended goal without colliding with anything [17].

In this paper, an innovative path-planning method was developed. The algorithm was divided into two main objectives. The first module deals with the optimization techniques employed in the grassfire algorithm and the implementation of the algorithm to the V-rep simulation, experimental analysis. The result of the sine cosine algorithm [17] was consistent with both outcomes at 5%, as previously explained. The primary aim of modeling and experimental results in this study was the reduction of range deviation by 4.6%. The total area under consideration was 350×300 cm. In the simulation, the distance covered was 270.97 cm, accomplished in a time was 21.58 s. During the experimentation phase, the distance covered was 283.87 cm, and the time was 22.59 s. It's important to note that the fixed grid structure of the graph contained 27% obstacles.

2 Graph Search Methods and Algorithm

With the help of one of the above-discussed graph creation techniques, the environment map has been transformed into a connection graph. Regardless of the map representation used, the objective of path planning is to locate the path that best satisfies the chosen optimization criteria (for example, the shortest path) between the start and the goal in the connection graph of the map.

$$f(n) - g(n) + \epsilon.h(n)$$

Where

f(n) = Expected total cost
g (n) = Path cost (accumulated cost of traveling from the start node to node n)
h (n) = Heuristic cost

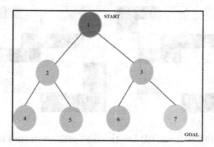

Fig. 1. Simple schematics of the grassfire algorithm.

The grassfire algorithm was found to identify points 1, 2, 3, 4, 5, 6, and 7 in this study. The starting point was 1, and the goal point was 7, and the shortest distance between these two points was found to be 1-3-7. This path represented the minimum distance. The remaining paths, 1-3-6-7, 1-2-4-7, and 1-2-5-7, represented the maximum distance points (Fig. 1).

Figures 2 and 3 that the grassfire algorithm was used for the shortest path motion. The main objective of the grassfire algorithm was to reach the start to goal point in the minimum distance path. The light blue color indicated the start to goal point identification for 1-3-6-7-9-12, and this distance was maximum compared to the other color-indicated lines. Next, the purple color indicated a point in the start-to-goal point identification for 1-3-6-8-11-12, and this distance was minimum compared to the other color indications of lines. Finally, the black color indicated a point in the start-to-goal point identification for 1-4-7-6-8-9-12, and this distance was maximum compared to the other color indications of lines.

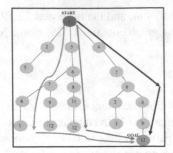

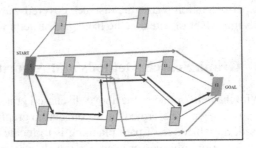

Fig. 2. Grassfire algorithm.   **Fig. 3.** Grassfire algorithm for the shortest path.

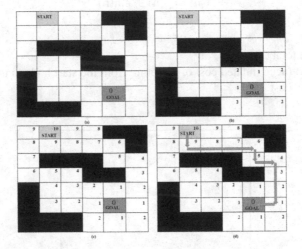

Fig. 4. (a), (b), (c), (d) Destination path from start to goal.

Graph structures 6 × 7 (6 columns and 7 rows) were used, with the vacant cells acting as the nodes and edges. The collection of nodes and edges made up graph G, which was composed of a collection of vertices (V) and a collection of edges (E) that connected pairs of vertices. The grid planning algorithm was known as the "grassfire algorithm." The objective was to create a route through the grid or graph from the beginning to the finish. Usually, there were many options for connecting two nodes. To start, the destination node was assigned a distance value of 0. Every iteration located and added a distance value of + 1 to all the unmarked nodes that were close to marked nodes. Next iteration adding from 1 + 1 = 2, Remaining iteration adding + 1 (for example 2 + 1 = 3) up to ninth iteration adding to + 1 (8 + 1 = 9). The distance value generated by the grassfire algorithm showed the shortest path between each node and the objective. Figure 4 (a), (b), (c), and (d) showed the route taken to reach the target. The start nodes were 10, and the goal node was 0. The block square box indicates to 27% of obstacles.

3 Optimization Techniques of Grassfire Algorithm

The three forces, the gravitational pull on the solution, and the environmental interaction between the solution and other grassfire algorithms were used to identify the grassfire algorithm in the optimized mathematical calculation to ascertain the Yi location of every solution.

$$Y_i = U_i + H_i + B_i \tag{1}$$

Y_i = Position of the robot
U_i = Environmental interaction between another grassfire
H_i = Gravity force of the solution
B_i = robot movement direction

The equation below represents the position of each solution in random numbers,

$$Y_i = C_1 U_i + C_2 H_i + C_3 B_i \tag{2}$$

C1, C2, C3 are random numbers of the array [0,1] model of each force in Eq. (1)

$$U_i = \sum_{j=1}^{n} U(e_{ij})\hat{e}_{ij} \quad i \neq j \tag{3}$$

$$e_{ij} = |y_j - y_i| \, and \, \hat{e}_{ij} = \frac{|y_j - y_i|}{e_{ij}}$$

where,
        U mentioned to the strength of two environmental forces (repulsion and attraction) e_{ij} the distance between i-th grassfire and j-th grassfire, $\hat{e}_{ij}$ unit vector (Fig. 5).

$$T = fc^{-s/m} - c^{-s} \tag{4}$$

Force of gravity

$$H_i = -h\,\hat{e}_h \tag{5}$$

where, $-h$ gravity constant, $\hat{e}_h$ *is unit vector towards target (or) goal*

$$B_i = U\,\hat{a}_w \tag{6}$$

where, U drift constant, $\hat{a}_w$ *is unit vector wind direction*

Grassfire Optimization Position
        In Eq. (3), (5), (6) is substitutes in Eq. (1).

$$Y_i = \sum_{j=1}^{n} U(e_{ij})\hat{e}_{ij} - h\,\hat{e}_h + U\,\hat{a}_w \quad i \neq j$$

$$Y_i = \sum_{j=1}^{n} U|y_j - y_i| \frac{|y_j - y_i|}{e_{ij}} - h\,\hat{e}_h + U\,\hat{a}_w \tag{7}$$

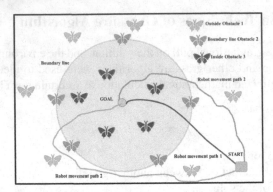

Fig. 5. Different pathway in robot movement.

They accomplish the optimization and best solution of the grassfire algorithm by fast reaching their comfort zone and swarming to the target site (global optimum), some modification in the previous Eq. (7).

$$Y_i^d = U \sum_{j=1}^{n} U\left[\frac{XB_d - YB_d}{2}\right] t \left|y_j^d - y_i^d\right| \frac{|y_j - y_i|}{e_{ij}} + best\ solution \qquad (8)$$

where $i \neq j$ U is mentioned to parameters co efficient, XB_d and YB_d Upper and lower bonds d-th dimensions.

$$d = d_{max} - iteration * \frac{d_{max} - d_{min}}{maxiteration} \qquad (9)$$

where d_{max} and d_{min} maximum and minimum values of d, max iteration means maximum iteration, d means number of iterations.

Algorithm 1: Grassfire Algorithm Optimization (GAO)

Parameter initialization: d_{max}, d_{min}, s and m.
Pseudocode initialization X_i (i=1,2, 3.........., n)
Select shortest distance is best (minimum distance)
While (current iteration (iter), maximum iteration (max iter)) **Do**
    Update d using equation (9)
    for grassfire **do**
        Distance between grassfire in the range [0,3.5]
        Grassfire opmization position equation (8)
        Current grassfire back goes in boundary line
    End
    Best solution find out
    Iter= iter+1
End
    Best solution

Algorithm 2: Pseudo code of the grass fire

STEP 1
For each node n in the graph
0 n * distance = infinity
STEP 2
Create an empty list
STEP 3
Goal distance = 0, add a goal to list
STEP 4
While the list is not empty
Let current = first node in the list, remove current from the list
STEP 5
For each node, n that is adjacent to current
If n. distance = Infinity
n. distance = current distance +1
add n to the back of the list
End.

---

4 Result and Discussion

4.1 Rep Simulation Analysis of Mobile Robot

The V Rep simulation was employed in the path planning in grassfire algorithm, carried out in simulation as to reach the starting and goal positions, various obstacles were present in the environment. Figure 6(a), (b), (c), (d), (e), (f) V-Rep simulation path in different obstacles in the environment. The Khepera-III robot was utilized for live experiments, while V-rep was employed for simulation. The step-by-step process of robot movement within six position points moved from the start to the goal point. The moving robot did not collide with the environment or experience friction on the surface for wheel movement. The V-rep simulation of the robot's movement was examined. It became evident that the path distance, which was calculated through simulation on the platform, deviated by less than 4.63%. This deviation fell within a respectable range. However, during the real-time experiment, various environmental factors, including surface friction and wheel slippage, caused the results to deviate.

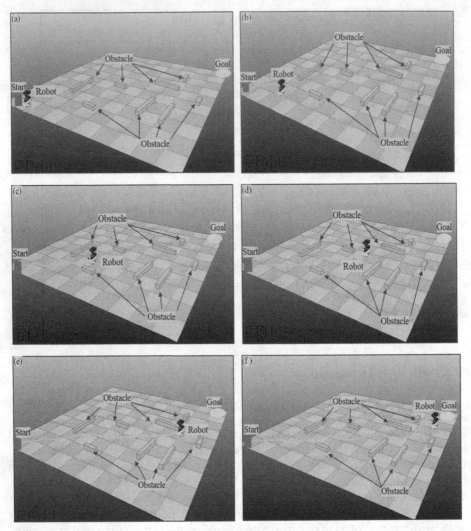

Fig. 6. (a), (b), (c), (d), (e), (f) V-Rep simulation path in different obstacles in the environment.

The number of runs was plotted on the X-axis, and the distance (cm) for both simulation and experimental data was represented on the Y-axis. The percentage difference value was calculated to be 4.63. The average simulation value was found to be 270.97 cm, while the experimental value was measured at 283.87 cm. Table 1 and Fig. 7 path distance travel in simulation and experimental process.

Table 1. Path distance in simulation and experimental analysis.

No of run	Simulation (cm)	Experimental (cm)	% Difference
1	269.91	282.76	4.65
2	270.96	284.54	4.86
3	272.51	285.41	4.62
4	264.46	280.36	5.83
5	268.49	284.43	5.76
6	270.51	284.83	5.15
7	272.81	282.61	3.52
8	270.81	280.72	3.59
9	271.52	282.39	3.92
10	269.81	286.36	5.95
11	274.57	286.39	4.21
12	275.61	285.71	3.59
Average	**270.97**	**283.87**	**4.63**

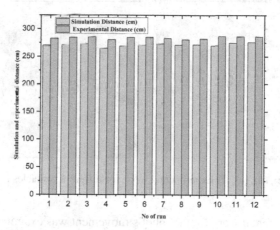

Fig. 7. Path distance travel in simulation and experimental process.

4.2 Experimental Result of Mobile Robot

In the experimental work, path planning to reach the starting and goal positions involved various obstacles in the environment. Figure 8(a), (b), (c), (d), (e), and (f) showed the experimental work for different obstacles in the environment. A thinker place stem mobile robot was utilized for live experiments, and an area of 350 × 300 cm with six position points in the same area was used. The step-by-step process of robot movement within the six position points was observed as it moved from the start to the goal point. The moving robot did not collide with the environment, despite the presence of different obstacles at the six position points of the robot's movement. In the research

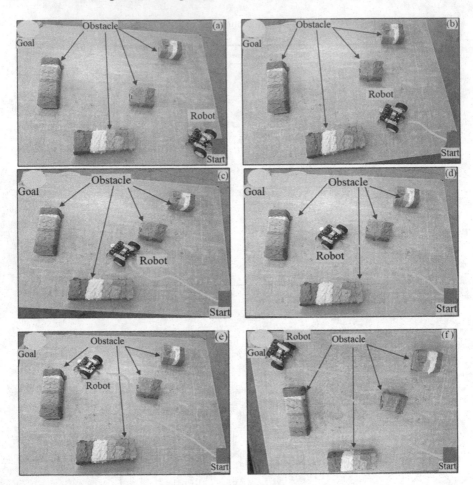

Fig. 8. (a), (b), (c), (d), (e), (f) Experimental work for different obstacles in the environment.

article, the experimental work for the robot's movement was examined. It was evident that the path distance and implementation of time, which were calculated through the experimental work platform, deviated by less than 4.57%. This deviation fell within a respectable range. However, during the real-time experiment, various environmental factors, including surface friction and wheel slippage, caused the results to deviate.

The number of runs was plotted on the X-axis, and the implementation time for both simulation and experimental data was represented on the Y-axis. The % difference value was calculated to be 4.57. The average simulation value was found to be 21.58 s, while the experimental value was measured at 22.59s. Table 2 and Fig. 9 Implementation time for simulation and experimental analysis.

Table 2. Implementation time for simulation and experimental analysis.

No of run	Simulation (sec)	Experimental (sec)	% Difference
1	21.64	22.40	3.45
2	21.32	22.53	5.51
3	21.36	22.62	5.72
4	21.38	22.40	4.65
5	21.42	22.53	5.05
6	21.44	22.72	5.79
7	21.47	22.85	6.22
8	21.49	22.45	4.36
9	21.82	22.42	2.71
10	21.85	22.68	3.72
11	21.87	22.73	3.85
12	21.97	22.83	3.83
Average	21.58	22.59	4.57

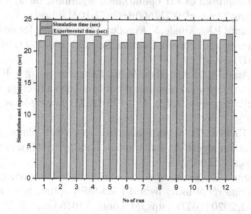

Fig. 9. Implementation time for simulation and experimental analysis.

5 Conclusion and Future Scope

In this article, the navigational control in path planning optimization of a mobile robot was one of the main objectives of the grassfire algorithm optimized methodologies. This method was utilized by the mobile robot from the target to the designated reached position. The GAO's main objective was to achieve closer proximity by adding one for each iteration, ultimately determining the best answer or the shortest distance to the destination. Khepera-III was employed to control and optimize navigation using the grassfire algorithm. Tables 1 and 2 presented the results from both the simulation platform and the experimental platform. The findings in the tables demonstrated a high level of agreement between the two platforms, with the observed variation being less than

4.6%. Additionally, the robot successfully accomplished the desired objective without any collisions.

Advanced grassfire algorithm optimization techniques were employed for path planning in navigational control using multiple mobile robots. The study involved the utilization of an advanced hybrid potential field path planning algorithm. A comparison was conducted with several other algorithms, including A*, RRT, Genetic Algorithm (GA), and the proposed algorithm.

References

1. Alabdalbari, A.A., Abed, I.A.: New robot path planning optimization using hybrid GWO-PSO algorithm. Bull. Electr. Eng. Inform. **11**(3), 1289–1296 (2022). https://doi.org/10.11591/eei.v11i3.3677
2. Fuad, M., Agustinah, T., Purwanto, D.: Collision avoidance of multi modal moving objects for mobile robot using hybrid velocity obstacles. Int. J. Intell. Eng. Syst. **13**(3), 407–421 (2020). https://doi.org/10.22266/IJIES2020.0630.37
3. Abdulsahebs, J.A., Kadhim, D.J.: Robot path planning in unknown environments with multi-objectives using an improved COOT optimization algorithm. Int. J. Intell. Eng. Syst. **15**(5), 548–565 (2022). https://doi.org/10.22266/ijies2022.1031.48
4. Dewang, H.S., Mohanty, P.K., Kundu, S.: A robust path planning for mobile robot using smart particle swarm optimization. Procedia Comput. Sci. **133**, 290–297 (2018). https://doi.org/10.1016/j.procs.2018.07.036
5. Kanoon, Z.E., Al-Araji, A.S., Abdullah, M.N.: Enhancement of cell decomposition path-planning algorithm for autonomous mobile robot based on an intelligent hybrid optimization method. Int. J. Intell. Eng. Syst. **15**(3), 161–175 (2022). https://doi.org/10.22266/ijies2022.0630.14
6. Abdulsaheb, J.A., Kadhim, D.J.: Multi-objective robot path planning using an improved hunter prey optimization algorithm. Int. J. Intell. Eng. Syst. **16**(2), 215–227 (2023). https://doi.org/10.22266/ijies2023.0430.18
7. Ajeil, F.H., Ibraheem, I.K., Sahib, M.A., Humaidi, A.J.: Multi-objective path planning of an autonomous mobile robot using hybrid PSO-MFB optimization algorithm. Appl. Soft Comput. J. **89**(June), 2020 (2022). https://doi.org/10.1016/j.asoc.2020.106076
8. Patle, B.K., Pandey, A., Parhi, D.R.K., Jagadeesh, A.J.D.T.: A review: on path planning strategies for navigation of mobile robot. Def. Technol. **15**(4), 582–606 (2019). https://doi.org/10.1016/j.dt.2019.04.011
9. Orozco-Rosas, U., Montiel, O., Sepúlveda, R.: Mobile robot path planning using membrane evolutionary artificial potential field. Appl. Soft Comput. J. **77**, 236–251 (2019). https://doi.org/10.1016/j.asoc.2019.01.036
10. Ou, J., Wang, M.: Path planning for omnidirectional wheeled mobile robot by improved ant colony optimization In: Chinese Control Conference CCC, vol. 2019-July, pp. 2668–2673 (2019). https://doi.org/10.23919/ChiCC.2019.8866228
11. Li, F., Fan, X., Hou, Z.: A firefly algorithm with self-adaptive population size for global path planning of mobile robot. IEEE Access **8**, 168951–168964 (2020). https://doi.org/10.1109/ACCESS.2020.3023999
12. Quan, Y., Ouyang, H., Zhang, C., Li, S., Gao, L.Q.: Mobile robot dynamic path planning based on self-adaptive harmony search algorithm and morphin algorithm. IEEE Access **9**, 102758–102769 (2021). https://doi.org/10.1109/ACCESS.2021.3098706

13. Abdullah, J.M., Ahmed, T.: Fitness dependent optimizer: inspired by the bee swarming reproductive process. IEEE Access **7**, 43473–43486 (2019). https://doi.org/10.1109/ACCESS.2019.2907012
14. Abed, I.A., Ali, M.M., Kadhim, A.A.A.: Using particle swarm optimization to solve test functions problems. Bull. Electr. Eng. Inform. **10**(6), 3422–3431 (2021). https://doi.org/10.11591/eei.v10i6.3244
15. Denk, M., Bickel, S., Steck, P., Götz, S., Völkl, H., Wartzack, S.: Generating digital twins for path-planning of autonomous robots and drones using constrained homotopic shrinking for 2D and 3D environment modeling. Appl. Sci. **13**(1), 105 (2023). https://doi.org/10.3390/app13010105
16. Kumar, S., Parhi, D.R., Muni, M.K., Pandey, K.K.: Optimal path search and control of mobile robot using hybridized sine-cosine algorithm and ant colony optimization technique. Ind. Robot. **47**(4), 535–545 (2020). https://doi.org/10.1108/IR-12-2019-0248
17. Kumar, S., Parhi, D.R., Kashyap, A.K., Muni, M.K., Dhal, P.R.: Navigational control and path optimization of mobile robot using updated sine–cosine algorithm in obscure environment. In: Acharya, S.K., Mishra, D.P. (eds.) Current Advances in Mechanical Engineering. LNME, pp. 989–996. Springer, Singapore (2021). https://doi.org/10.1007/978-981-33-4795-3_91
18. Zhong, X., Zhou, Y., Liu, H.: Design and recognition of artificial landmarks for reliable indoor self-localization of mobile robots. Int. J. Adv. Robot. Syst. **14**(1) (2017). https://doi.org/10.1177/1729881417693489

A Novel Approach for Suggestions on Law Based Problems

Bhargav Vyas[1([✉])] and Jeegar Trivedi[2]

[1] MCA Department, CHARUSAT University, Changa, Gujarat, India
bhargavvyas2793@gmail.com
[2] Computer Center, M.S. University, Vadodara, Gujarat, India

Abstract. The focus of this paper is all about using artificial intelligence methods for the legal problems and getting accurate results. The implementation of different AI technologies clubbed with the law domain based expert system is discussed in detail. The architecture, implementation and results are discussed in detail for getting a clear view. The experiments are carried out using the legal concepts of IPC (Indian Penal Code) and compared. The findings of the paper are different and novel which can be interesting for the experts of legal domain. The different methods which can provide similar results are also discussed and the comparative analysis is done. The proposed model discussed in the paper provides significant results in terms of accuracy.

Keywords: Fuzzy Logic · Expert System · Membership Functions · Linguistic variables · Association rule mining

1 Introduction

The domain specific knowledge is given primary importance in recent times. The person can achieve knowledge regarding particular domain by learning, understanding and applying, but cannot have knowledge of multiple domains at the same time. The same example can be considered for common people, who do not possess any knowledge regarding the law domain which is intended to protect their rights. Many times it may happen that the person involved in any legal situation may be unaware about the consequences of it. The major problem may arise is getting cheated by the people having good knowledge related to the legal domain. The introduction of artificial intelligence in the legal domain can solve a great of this problem. The discussion is about a well-defined architecture for legal expert system. There are many different sub categories in the law based domain which need individual solution for specific category. The theoretical problems can be easily solved by applying logic but when it comes to real world problems, there is a requirement of applying algorithm.

Fuzzy logic can be used for controlling complex systems when enough amount of knowledge is available. There are many systems which are using fuzzy logic as a solution, no matter if it is commercial or non-commercial. Fuzzy logic has a wide use but in this case it will be considered only for legal expert system [1]. The fuzziness is defined by

© The Author(s), under exclusive license to Springer Nature Switzerland AG 2024
K. K. Patel et al. (Eds.): icSoftComp 2023, CCIS 2031, pp. 190–200, 2024.
https://doi.org/10.1007/978-3-031-53728-8_15

the membership function by identifying the degree of membership. The membership is identified by the set of rules written as "if-then". It can be considered as most suitable for the domain as it can deal with the applications where the data may be imperfect. The crisp values are associated with some degree of membership which can be between 0 and 1 [2]. The identification of the membership function can be done using some linguistic labels. The labels can be considered as some descriptive parameter which can be used for dividing the area of universe. Each divided area can be added with some range of values and when the new value is entered in the system, it will be verified with the ranges to identify the corresponding linguistic label [3]. There are few basic concepts like support, core and height which are much important to be understood before defining the membership function [4]. The members can have different values associated with them between 0 and 1, if any member has value greater than 0 can be considered as support. The members whose degree is equal to 1 are called as core elements and the maximum degree of membership is known as the height of the membership function [5].

The fuzzy rule based expert system made for the legal domain will be a combination of modules like fuzzification unit, defuzzification unit, knowledge base and the decision making unit. The rules made for measuring the degree of membership are responsible for the accuracy provided by the system. The importance of using the technology is human understandable input and output to and from the system respectively. The rule base may comprise of different content collected related to domain, which may be from internet, book, experience, etc. [6]. The flow of the system for getting accurate suggestions will be as shown below [7, 8],

- Crisp input will be provided to the system
- The fuzzification unit will Fuzzify the crisp input
- The decision making unit will use the fuzzified input with the knowledge base for finding the degree of membership
- The defuzzification unit will help in getting the actual output back by converting the fuzzy output to the crisp output.

Association rule mining is also applied to the system which helps in filtering the huge amount of data available. Interesting patterns related to the input are identified and stored for faster execution of similar data [9]. It is proven technique which is used for the efficient and accurate results in case of huge database, same as we have in the law based domain. This is an iterative technique which should be continued until accurate result is found [10].

1.1 Objectives

The people have very less knowledge related to legal domain. The research will target the sections between IPC 300 to 400. The research will not only help the layman's but also the people who are new in practicing law. New rules are added frequently and most of the people are unaware about that embedment's. It is difficult task to remember all the rules and so the work also helps in finding related rules and accessing them. The work is used for giving suggestions which can be understood by all. The suggestions can help in improvement of the decisions taken regarding the legal problem.

1.2 Organization

Section 2 shows the related work done throughout world in the same technology or in the same domain. Section 3 shows the novel framework which is used for solving law based problems of IPC. Section 4 shows the technical background and the actual implementation of the framework. Section 5 describes the results obtained using the competitive technologies and the key points of benefit of new framework. Section 6 shows the future work which can be carried out in the same domain and technology.

2 Related Work

There are many systems which are working with law based domain. Most efforts are given for covering the systems which are closely related to the work. Some systems are close to the domain of work and some are related to the technique used in the work. The first related work is of a system developed by John Campbell and Kamalendu Pal which was named as ASHD-II. It was the case based and rule based reasoning method used for matrimonial cases in UK. Generally, it was used to settle the property related issues. The issue with the system was it selected the suitability of the method i.e. case based reasoning or rule based reasoning randomly and provided with the judgement [11].

The second system in discussion is CHIRON which was developed by Kathryn E. Sanders which was again working on rule based and case based reasoning. The system is used for taxation related rules and problems of USA. In this system the data is passed to rule based planner and then to the case based planner for testing of the results. It is a cyclic process and continue until best result is obtained. The only problem id iteration of the process until best result obtained which is time consuming [12].

The third nearest system is JUDGE which was designed by William M Bain which worked with the rule based reasoning in a way as if it is applying the case based reasoning. In this system, the generation of new rules is started by traversing through the older cases. The only problem will arise when the new case will be given to the system where there are no similar cases in history of the system. Under such scenario the system will not be able to give precise output [13].

The next system us TAXMAN which is working with the tax based laws of USA and is developed by McCarty Throne. The major focus of this system is corporate tax of USA and the problems related to that. The method used while taking input is forward chaining and the expands it to greater level and the result is generated. For checking the accuracy of the result, backward chaining technique is used which can check if all the conditions are satisfied or not. No accuracy can be expected from the system after completion of forward and backward chaining [14].

Another system developed by Thoen Walter and Yao-Hua Tan which was named as INCAS works for the legal domain working on contracts between the customers and ecommerce websites. It may not provide the same quality of results for all the online websites as the terms and conditions may change in every case [15].

Another system was proposed by Burkhard which was named ZOMBAIS. The name itself shows that the system is used for the rules related to the documents which are beyond the grave. In this case the system takes input as a document and process the document for understanding the legal reasoning. The major problem with the system is it will not understand the exact feeling of the person who has written the will and may lead to an unfair decision which may not be exactly as per the will [16].

The systems were near to the legal domain but none of them is working on the same domain which is proposed. The next is the discussion about the systems working on similar technology.

The first system is for diabetes diagnosis which is a common disease for mid aged people in world. Diabetes have a lot of variations in terms of parameters of detection. The decision control tree is used for finding the exact match of the rules. The model used for detecting the most common disease under discussion is Mamdani [17].

The next is hybrid fuzzy approach for detection of Parkinson which is again a disease where there is lot of imprecision in detection parameters. The rules should be minimized to get faster results which may lead to inaccuracy. In the first part the fuzzy rules are generated automatically using the input parameters and in the second stage the fuzzy rules are applied to detect the final result [18].

The system in discussion is using soft set and fuzzy expert system for health care and medical issues identification. The medical science issues identification is a great challenge as it has a lot of complications involved and may vary person to person. The soft set is used for solving the problem by applying the rules and making the decision for particular medical issue [19].

There are many different system working on the concept of fuzzy logic or a combination of other techniques. The systems discussed are very close to the domain and techniques used in current work. There are many law based systems in the world but none of them is being used by common people. It is generally used by the people who are pursuing the legal carrier. There is again no generalized system which can make suggestions for all the type of domains which consists of rules. The systems which are discussed have advantages as well as disadvantages but none of the system is working in the same way as the proposed work is doing. The proposed work system is for the people having less legal knowledge and can get best legal suggestions without being cheated by the experts.

3 Proposed Model

The model will display all the components and the flow of the system for getting the best possible suggestions. The user will need to give domain specific input as per the requirement and will get the best possible output for the same. Suppose in the current scenario the system will take input as severity of crime, the definition and the possible description and will process for the best suggestion being provided (Fig. 1).

Proposed Novel Framework Steps:

Step1: Select the domain and input for which the approach is to be applied
Step2: Take the input from the user in required format
  $input = takeUserInput()$
Step3: Decide the antecedents and the consequents to the system using association rule mining
  $association\_rules = mineAssoRules(array\_atecedent, array, consequent)$

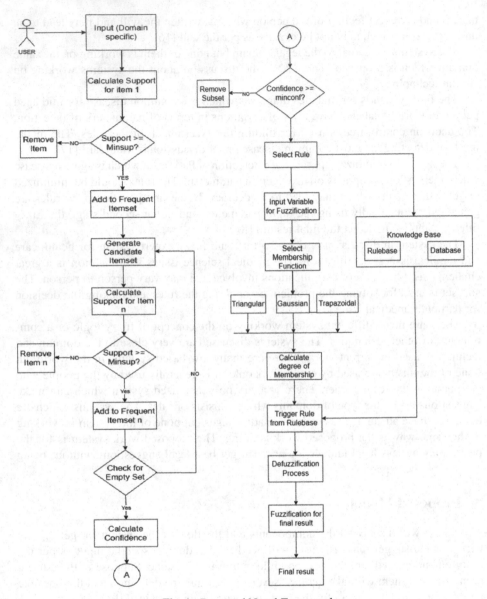

Fig. 1. Proposed Novel Framework

Step4: Generate the fuzzy set list which can be applied on the selected domain
    fuzzySetList = generateFuzzySetList(input)
Step5: Select the best membership function to be applied on the input data
    membershipFunction = selectMembershipFunction(input)
Step6: Use the MAX operator for combining the value from membership function and
the association rule mining
    result = applyMaxOperator(membershipFunction, association_rules)

Step7: Store the result and apply AND operator to merge both the results

$mergedResult = applyAndOperator(result, fuzzy(result))$

Step8: The defuzzification method need to be used for converting the fuzzy result into a crisp value.

$crispValue = defuzzify(mergedResult)$

$fuzzyAdvice = fuzzify(crispvalue)$

Step 9: Final crisp value will be fuzzified again for the generation of advice

$output = defuzzify(fuzzyAdvice)$

4 Methodology

The actual implementation of the new approach is done step by step in this section. The fuzzifier is used for converting the crisp input from the user into fuzzy values and to check the degree of membership. The most important part in any fuzzy logic system is the rule base which can change the final result from the system. The precision of the rules can define the precision of the result. The rules are divided into two parts which can be defined as antecedents and consequents. The degree of similarity is measured by the membership functions by comparing the input with the rules. The rules can be used as and when required from the rulebase and upgraded if there are changed according to the domain. After the extraction of rules, the final step is the defuzzification of the fuzzy output to get the final result. The detailed description of the steps is as defined below,

Step1: Provide the parameters of Input

The developed approach will accept a list of input from the user which will be linguistic in type. For example, in the case of legal domain the input will be the values of severity, definition and the description. All the variables can have multiple values possible which lead to a combination of huge number of option for input. For solving the same problem and getting faster input from the user the association rule mining technique can be applied to get the frequent set of input and provide it to the user by default.

Step2: Supplying input parameter to Association Rule Mining

The input related to particular domain will be added for the association rule mining process and the relationship between them can be easily identified. There are many different methods being used for the association rule mining but in this particular case the Apriori algorithm is considered. The association rule can be stored so that the next time same set of input can be fetched. There are three input possible and so in such case we can create the candidate tables by setting some support value. The candidate tables are created until the support count limit is reached and no further combination is possible for the input set. The confidence value need to be calculated and in case of the maximum value i.e. 1, the rule should be created. The value 1 of confidence means that if two values of input parameter is used together and if there is a rule made for the same combination than the third value will be selected automatically.

Step 3: Linguistic Variable Declaration

According to the linguistic values the input variables are fixed and using that we can further fix other variables namely bailable, court type, cognizable and rule type. The values can be associated with the linguistic variables according to the domain used for getting the suggestion.

Step4: The fuzzification process

The process of converting the crisp input to fuzzy ones using the membership function is known as the fuzzification process. Generally, single membership function can be used but there are chances in certain case where more than one membership function can be used. In this case only one membership function i.e. triangular is used. The fuzzy numbers which are associated with each and every variable taken as input from the user can be then added to the fuzzy set. In this case three values will be associated with each linguistic variable as triangular membership function is used. After all the process of fuzzification all the inputs will be merged into a single fuzzy set.

Step5: The fuzzy inference process

The if then rules can be generated which in general will look as below,
IF cognizable is value, bailable is value, court type is value THAN the result is generated. There can be more than one rule for each condition and all of them can be checked as and when required. In our case more than 2500 rules are created which can be used for getting the best and accurate results of the input for particular legal domain. The rule can be described as shown below,

```
If(Cognizable == "value")
{
    If(Bailable == "value")
    {
        If(Court_type == "value")
        {
            Result = "advice"
        }
    }
}
```

Step6: The defuzzification process

The process of converting the fuzzy values to the crisp values is known as defuzzification. In the current example, the centroid method for defuzzification is used which helps in getting the actual result in terms of advice. The demonstrated example shows that the best result is possible for particular domain. The output generated will be very easy to understand and in a simple language for which no legal knowledge is required.

5 Result and Discussion

For validating the success of the model, huge amount of experiments is conducted. Huge amount of different inputs and test cases are considered. The selection of fuzzy sets and integration with the association rules into a cohesive system is new component with the research work. The application of novel algorithm performs well in the legal domain. The parameters are well tuned for the best results according to the legal domain. The work is compared for the results by comparing it with different techniques as well as different systems available in world. The proposed model is working on the IPC rules starting from 301 to 400 which are the laws applicable when affecting the human body. The system under testing consists of more than 2500 rules which can be used in all the different types of experiments. In the experiments carried out, mainly 7 cases are discussed which shows that the experiments are satisfactory. Out of total number of experiments carried out, the system is providing 98.5% accurate results when compared with the legal reference books. There are many people who are related to the legal domain and have tested the system for accuracy and many have also provided with positive testimonials. In particular domain, around 100 cases having diffcrent input are given to the system and compared the results. Few cases where the deviating results are obtained are discussed. There is comparison of 3 techniques like the proposed model, Mamdani and Sugeno. The experiments are carried out using matlab. In the result comparing three techniques, in many cases two techniques are providing same result and any one may be deviating or in some cases all the three have different results generated.

The Mamdani model was used for experiment using Matlab. Triangular membership function was used for getting the suggestions as shown below (Fig. 2),

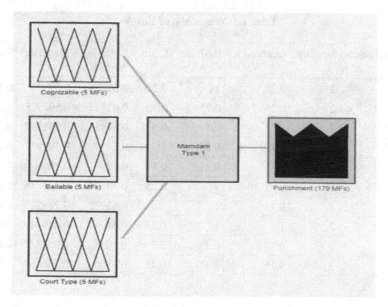

Fig. 2. Mamdani Fuzzy Inference System

The next experiment was carried out using the Sugeno model and again the same triangular membership function was used for carrying out the experiment as shown below (Fig. 3 and Tables 1, 2),

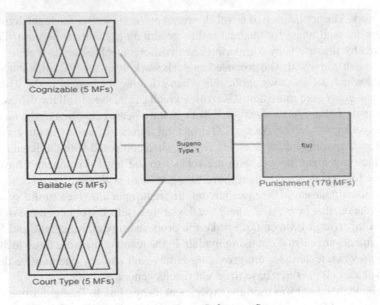

Fig. 3. Sugeno Fuzzy Inference System

Table 1. Comparison of Results

Sr. No.	Definition	Severity	Cognizable	Bailable	Court	Novel Approach	Mamdani	Sugeno
1	3	2	0.25	1	0.25	0.60211	0.59	0.58809
2	2	2	0.25	0.25	0.75	0.61797	0.592978	0.58447
3	4	3	0.75	0.1	0.5	0.67857	0.658597	0.64316
4	3	3	0.75	0.25	1	0.2	0.182977	0.2
5	3	1	0.75	0.1	0.5	0.12	0.106	0.12
6	7	3	0.75	1	0.25	0.39325	0.371718	0.39325
7	7	3	0.75	1	0.5	0.82022	0.789225	0.82022
8	2	2	1	1	0.5	0.64044	0.619449	0.64044
9	8	3	0.75	0.1	0.5	0.09333	0.09333	0.05098
10	3	3	0.75	1	0.75	0.98666	0.986667	0.94769

Table 2. Comparison of Technologies

Sr. No.	Method Used	MF's	Accuracy	Remarks
1	Novel Approach	Triangular	98.6%	Best for domain specific advice, best suited for closed domain expert system and provides high accuracy in results
2	Mamdani	Triangular	91.3%	Suitable for rough data set and generic FIS but lacks accuracy in closed environment
3	Sugeno	Linear	88.1%	Suitable for fixed data set, but lacks in accuracy for degree of membership in output

6 Conclusion and Future Scope

A novel fuzzy algorithm is developed for the legal advices by combining association rule mining and fuzzy inference. The selection of fuzzy set in an unexplored domain is also a unique thing which provides the best results when compared. It is useful for all the people associated with legal domain and need frequent advices. The lawyers and judges can mostly have the benefit of the work as it can help in faster evaluation of cases.

The proposed framework can be easily expanded by adding the rules of particular domain. The linguistic variables can also be increased for better accuracy of the system. The previously added suggestions and results can be stored for faster evaluation of similar case the next time. With the increase in rules and linguistic variables for any domain can increase the accuracy of the work. There are many ways to get input to the system like voice based input. Other machine learning techniques can also be added in the existing framework for better results.

References

1. Negnevitsky, M.: Artificial Intelligence: A Guide to Intelligent Systems, pp. 25–48 (2001)
2. Zadeh, L.A.: Fuzzy sets. Inf. Control. **8**(3), 338–353 (1965)
3. Lotfi, A., Tsoi, A.C.: Importance of membership functions: a comparative study on different learning methods for fuzzy inference systems. In: Fuzzy Systems, 1994. IEEE World Congress on Computational Intelligence, Massachusetts (1994)
4. Dombi, J.: Membership function as an evaluation. Fuzzy Sets Syst. **35**(1), 1–21 (1990)
5. Cheema, J.S., Singh, I.: Fuzzy Systems. Khanna Book Publishing Co.(P) Ltd., Delhi (2011)
6. Aly, S., Vrana, I.: Toward efficient modeling of fuzzy expert systems: a survey. Agric. Econ. **52**(10), 456–460 (2018)
7. Amaral, J.F.M., Tanscheit, R., Aur, M.: Evolutionary fuzzy system design and implementation, Singapore (2002)
8. Setnes, M., Babu, R., Verbruggen, H.B.: Rule-based modeling: precision and transparency. IEEE Trans. Syst. Man Cybern. Part C (Appl. Rev.) **28**(1), 165–169 (1998)
9. Ziauddin, Z., Kamal, S., Ijaz, M.: Research on association rule mining. Adv. Comput. Math. Appl. **2**(1), 226–236 (2012)

10. Shoemaker, C.A., Ruiz, C.: Association rule mining algorithms for set-valued data. In: Intelligent Data Engineering and Automated Learning, 4th International Conference, Hong Kong (2003)
11. Campbell, J., Pal, K.: ASHSD-II: a computational model for litigation support. Expert. Syst. **15**, 169–181 (2008)
12. Sanders, K.E.: Representing and reasoning about open-textured predicates. In: ICAIL 1991: Proceedings of the 3rd International Conference on Artificial Intelligence and Law, Oxford (1991)
13. Popple, J.: A Pragmatic Legal Expert System, 1 edn. Dartmouth Publishing Company Limited, Hants (1996)
14. McCarty, T.L.: Reflections on TAXMAN: an experiment in artificial intelligence and legal reasoning. Harvard Law Rev. 837–893 (1977)
15. Thoen, W., Tan, Y.: INCAS: a legal expert system for contract terms in electronic commerce. Decis. Support Syst. 389–411 (2000)
16. Schafer, B.: ZombAIs: legal expert systems as representatives "beyond the grave". Scripted **7**(2), 384–393 (2010)
17. Perumalsamy, D., Palanigurupackiam, N.: An intelligent fuzzy inference rule-based expert recommendation system for predictive diabetes diagnosis. Int. J. Imaging Syst. Technol. (2022)
18. Nilashi, M., Shahmoradi, L., Ahmadi, H., Ibrahim, O.: A hybrid intelligent system for the prediction of Parkinson's Disease progression using machine learning techniques. Biocybernetics Biomed. Eng. 22–26 (2017)
19. Salleh, A.R., Bashir, M.: Fuzzy parameterized soft expert set. Abstract Appl. Anal. 57–63 (2012)

Performance Evaluation of Service Broker Policies in Cloud Computing Environment Using Round Robin

Tanishka Hemant Chopra and Prathamesh Vijay Lahande(✉)

Symbiosis Institute of Computer Studies and Research, Symbiosis International (Deemed University), Pune, India
prathamesh.lahande@sicsr.ac.in

Abstract. With its ability to provide scalable and practical solutions for various applications, the cloud computing platform has emerged as a pillar of Information Technology. The **R**esource **A**llocation **A**lgorithms (RAA) of the cloud computing platform and its **S**ervice **B**roker **P**olicies (SBP) plays an impactful significance for its performance. Hence, it becomes essential to examine these SBPs used by the RAA. The primary aim of this research paper is experimentally examining the SBPs of the cloud, namely **C**losest **D**ata **C**entre (CDC), **Opti**mized **R**esponse **T**ime (OptiResTime), and **D**ynamically **R**econfigured (RD) using the RAA Round – Robin (RR). To do so, this paper includes experimenting with a cloud simulation platform and computing tasks in the cloud DCs in various scenarios. The performance parameters used for the study are **Overall R**esponse **T**ime (OvrallResTime) and **D**ata **C**entre **P**rocessing **T**ime (DCPT) measured in milliseconds (ms). The experimental results convey that the SBPs CDC, OptiResTime, and RD take an average OvrallResTime of 1310.68 ms, 1310.92 ms, and 8226.72 ms, respectively. Concerning DCPT, the SBPs CDC, OptiResTime, and RD take an average of 1010.71 ms, 1010.66 ms, and 7925.28 ms, respectively. Hence, the SBP CDC outperforms other SBPs in terms of OvrallResTime, and the SBP OptiResTime outperforms other SBPs in terms of DCPT. To enhance SBPs and maximize cloud results, the cloud needs to be embedded with an external intelligence mechanism. Therefore, this paper also presents a Reinforcement Learning model to enhance these SBPs and provide **Q**uality **o**f **S**ervice (QoS).

Keywords: Cloud Computing · Performance · Reinforcement Learning · Resource Allocation · Service Broker Policy

1 Introduction

Cloud computing platform has developed into the foundation of contemporary information technology due to its simplicity, easy scalability, and accessibility. Optimizing cloud computing resources is critical in an era characterized by the digital transformation of companies and rising reliance on cloud-based services. Efficient **R**esource **A**llocation **A**lgorithms (RAA) using the appropriate **S**ervice **B**roker **P**olicies (SBP) is critical in

K. K. Patel et al. (Eds.): icSoftComp 2023, CCIS 2031, pp. 201–213, 2024.
https://doi.org/10.1007/978-3-031-53728-8_16

guaranteeing the ideal **Quality of Service** (QoS) for end-users in this changing environment. In various challenging circumstances, the choice and assessment of these RAA with their respective SBPs have emerged as crucial determinants of cloud performance [2, 3]. The primary purpose of this research is to evaluate the performance of the cloud's SBP using the **Round-Robin** (RR) **Resource Allocation Algorithm** (RAA) through experimentations [17]. The SBPs considered for this study are the **Closest Data Center** (CDC), **Optimize Response Time** (OptiResTime), and **Reconfigure Dynamically** (RD). The performance parameters considered for this study are **Overall Response Time** (Ovrall-ResTime) and **Data Centre Processing Time** (DCPT) measured in **milliseconds** (ms). This research paper uses the RAA RR to compare the cloud's SBPs and their behavior concerning performance parameters. For this comparison, the authors of this paper have experimented in the cloud's simulation environment, where tasks were computed using the said SBPs considering the RR RAA in several scenarios. The experimental results convey the ideal SBP to enhance cloud computing performance. Lastly, the Machine Learning (ML) sub-domain of Reinforcement Learning (RL) model is suggested to add intelligence to the SBPs to increase the overall cloud performance [11, 12, 40].

Figure 1 represents the entire flowchart of the experiment.

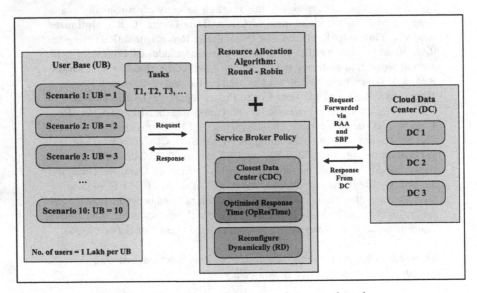

Fig. 1. Flow diagram for the experiment conducted.

The rest of the paper is organized as follows:

- Section 2 represents the detailed literature review;
- Section 3 represents the experimental design;
- Section 4 describes the results and their implications;
- Section 5 illustrates the cumulative results and RL to improve Cloud Performance,
- Section 6 includes the conclusion.

2 Literature Review

This section represents the detailed literature review conducted on the existing study. Several researchers have focused on solving the issues of cloud computing platforms and provided various methodologies and strategies. The researchers in this paper have tried to optimize the datacenter selection criteria for cloud computing and offered a novel method employing differential evolution algorithms [2]. This study uses a service broker strategy to investigate data center selection in cloud computing, addressing effective resource allocation and administration in cloud systems [4]. This study thoroughly examines QoS-aware load balancing (LB) approaches in generic and specialized fog deployment scenarios, providing valuable insights into optimizing Quality of Service (QoS) for fog computing environments [3]. The study introduces eMRA, a multi-optimization-based resource allocation technique for infrastructure clouds [9].

This research conducts a cognitive study of cloud computing assignment algorithms, providing insights into their effectiveness and success in managing resource allocation for shifted workloads [19]. This study provides a successful service broker policy [1] for intra-datacenter LB to boost resource allocation and enhance performance in data center environments [33]. This study delivers an efficient resource-sharing strategy for dwellings in intelligent grids that use fog and cloud computing technologies to enhance energy management and resource utilization [29].

This study looks into the issues and challenges of using the cloud, analyzing significant concerns and problems in this quickly growing technology field [38]. This study provides insights into their many uses and methodologies within the cloud computing ecosystem [14]. The current research analyses dynamic resource provisioning in cloud computing, emphasizing LB and service broker policies to optimize resource allocation in response to changing workloads [22]. This paper outlines a priority-based approach for enhancing cloud computing data center selection with broadened asset allocation techniques based on distinct goals and requirements [23]. The article assesses the performance of load-balancing algorithms in conjunction with various service broker policies for optimizing cloud computing systems [5]. This study describes a novel way to optimize intelligent grid management that combines cloud-fog layer technology with the Honeybee Mating Optimization Algorithm [6]. This review paper focuses on cloudlet allocation policies, presenting a synopsis of existing methodologies and their usefulness in improving resource allocation in cloudlet-based environments [27]. A self-learning optimizer is introduced in this dissertation to improve job scheduling in various fields by applying predictive capabilities and efficiency techniques [30]. The consequence of SBPs and algorithmic methods for LB on the functioning of large-scale internet applications in cloud data centers is looked into in the present study [31]. This paper addresses the function of multi-cloud service brokers in determining the best data center within a cloud environment, emphasizing their importance in optimizing resource allocation and improving cloud performance [7]. This work suggests a modified differential evolution strategy for optimizing cloud data center selection, with potential advances in the percentage of resources and overall cloud performance [10]. A multi-cloud resource brokering system was created for enhancing bioinformatics workflows [24].

This work aims to improve the efficiency and performance of cloud simulations by introducing unique service brokers and load-balancing algorithms tailored for CloudSim-based visual modeling [8]. Using the CloudAnalyst simulator this study investigates the worth of various algorithms and scheduling techniques for boosting computing efficiency [18]. The present piece evaluates and explores service aggregator algorithms beneath the Cloud-Analyst framework, highlighting their efficacy in optimizing resource allocation and enhancing general cloud system performance [15]. This study offers a LB for power utilization in intelligent grids using cloud computing, improving efficiency in handling resources within the grid infrastructure [34]. This paper contains an overview and viewpoint on the usage of cloud computing, which includes views and comments from several authors on multiple facets of the new technology and its impact on the computing environment [37]. The paper analyses hybrid cloud computing as a cost-effective solution to cloud interoperability concerns [13]. The current research offers a RL method to optimize dynamic resource allocation, providing a way for optimizing the allocation of resources [39]. This study analyses the performance of SBPs within the Cloud Analyst framework [21].

This study recommends a cost-effective service broker strategy for data center allocation in the IaaS cloud paradigm, focusing on effective resource allocation while minimizing costs [16]. This study analyses load-balancing algorithms and service broker guidelines in cloud computing, considering multiple user grouping guidelines to maximize resource allocation and improve cloud performance [26]. This study presents an optimized SBP for fog/cloud locations leveraging the differential evolution process to improve resource allocation and workload management efficiency and effectiveness [36]. The present research investigates the role of the closest data center as the SBP in cloud computing situations by analysing the effect of diverse device characteristics inside a round-robin-based balance of load algorithm [25]. The research presented here describes a cloud and fog-based brilliant grid setting that allows efficient energy management intending to boost the environmental responsibility and success of energy distribution and consumption [32]. This study addresses cloud and fog computing use for intelligent grid supervisors, focusing on improving power effectiveness and flexibility in modern grid systems [28]. The paper offers a novel SBP in cloud computing that optimizes both cost and timing of response [20]. This study describes a dynamic cost-load aware service broker LB technique designed for virtualized environments. It emphasizes efficient resource allocation when speaking of cost and load to reduce the usage of resources in virtual environments [35].

3 Experimental Design

This section represents the experimental design of the conducted experiment. The Cloud-Analyst cloud simulation environment was utilized to conduct the experiment. The resource allocation algorithm (RAA) Round – Robin (RR) was used to manage the cloud's resources. A user base is considered which possesses a huge number of users, which submit the immense number of tasks to the cloud DC for computations. To process the tasks, the Service Broker Policies (SBP) used are as follows:

- **Service Broker Policy (SBP) - Closest Data Center (CDC):** The SBP CDC prioritizes the allocation of tasks to the nearest DC to minimize latencies.
- **Service Broker Policy (SBP) - Optimize Response Time (OptiResTime):** The SBP OptiResTime aims to lower the response time and accordingly assigns the computational tasks.
- **Service Broker Policy (SBP) - Reconfigure Dynamically (RD):** The SBP RD represents a policy which adapts the resource allocation based on the changing dynamic conditions of the cloud.

The performance parameters considered for the study and comparison of these SBPs are as follows:

- **Performance Parameter - Overall Response Time (OvrallResTime):** The total amount of time, measured in milliseconds (ms), required to compute all the submitted tasks from its time of submission until it gets its first response.
- **Performance Parameter - Data Centre Processing Time (DCPT):** The total amount of time, measured in milliseconds (ms), required to compute all the tasks at the cloud DC.

The experiment was conducted in ten scenarios, where each phase differed from one another with its number of users in the user base. The experimental phases can be represented as follows:

Table 1. Experimental Scenarios.

Scenario No.	User Base	Total no. of Users	Scenario No.	User Base	Total no. of Users
1	1	1,00,000	6	6	6,00,000
2	2	2,00,000	7	7	7,00,000
3	3	3,00,000	8	8	8,00,000
4	4	4,00,000	9	9	9,00,000
5	5	5,00,000	10	10	10,00,000

Table 1 depicts that the number of users were 1 lac in the first scenario, and increments with one lakh in each scenario, until the tenth scenario where 10 lac users were used. In each scenario, the cloud DC is tested with the RAA RR with the said SBPs and the results of the SBP are compared with the said performance metrics.

4 Results and Implications

This section represents the detailed results and implications of the experiment conducted, represented in three sub-sections: sub-Sect. 4.1 represents the results with the SBP CDC; sub-Sect. 4.2 represents the results with the SBP OptiResTime; sub-Sect. 4.3 represents the results with the SBP RD. Each sub-section has a detailed comparison with respect to both the performance parameters OvrallResTime and DCPT.

4.1 Results with Respect to Closest Data Centre (CDC)

This sub-section represents the experimental results with respect to the Service Broker Policy (SBP) Closest Data Centre (CDC) using the Resource Allocation Algorithm (RAA) Round – Robin (RR). Figure 2 and 3 represents the graph of SBP CDC using RAA RR for the performance parameter Overall Response Time (OvrallResTime) and Data Centre Process Time (DCPT), respectively.

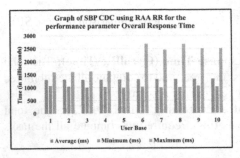

Fig. 2. Graph of SBP CDC using RAA RR for the performance parameter Overall Response Time (OvrallResTime).

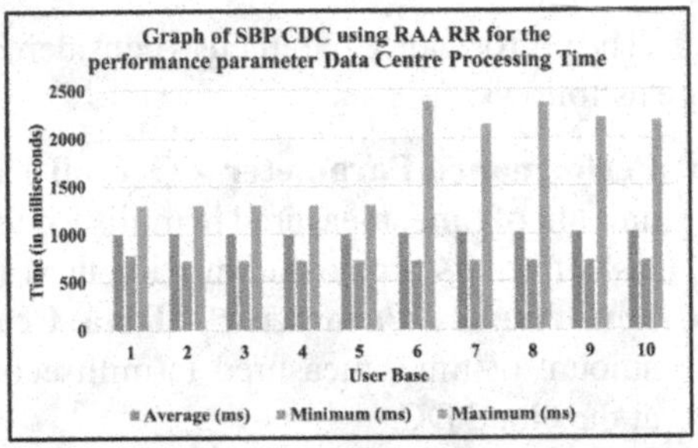

Fig. 3. Graph of SBP CDC using RAA RR for the performance parameter Data Centre Processing Time (DCPT)

Table 2 represents the experimental results concerning SBP CDC using the RAA RR.

Table 2. Experimental Results concerning SBP CDC using the RAA RR.

Scenario No.	Overall Response Time (OvrallResTime) (in milliseconds)			Data Centre Processing Time (DCPT) (in milliseconds)		
	Average (ms)	Minimum (ms)	Maximum (ms)	Average (ms)	Minimum (ms)	Maximum (ms)
1	1309.64	1075.52	1598.53	1009.94	780.51	1300.51
2	1311.84	1042.52	1609.02	1012.26	720.51	1300.51
3	1304.39	1005.01	1622.53	1005.31	720.51	1300.52
4	1300.23	1021.51	1630.03	1000.16	720.01	1300.01
5	1301.97	984.01	1577.52	1001.52	720.51	1300.51
6	1309.7	1027.51	2686.07	1009.02	720.51	2380.06
7	1313.91	988.52	2453.58	1013.91	720.51	2141.57
8	1315.89	1006.51	2672.5	1015.86	720.51	2365.49
9	1318.68	1056.01	2498.04	1018.52	720.51	2207.51
10	1320.52	1026.03	2502.58	1020.6	720.51	2180.57

From Table 2, we can observe that the

- Cumulative Average for OvrallResTime = 1310.68 ms
- Cumulative Average DCPT = 1010.71 ms

4.2 Results with Respect to Optimize Response Time (OptiResTime)

This sub-section represents the experimental results with respect to the Service Broker Policy (SBP) Optimize Response Time (OptiResTime) using the Resource Allocation Algorithm (RAA) Round – Robin (RR). Figure 2 and 3 represents the graph of SBP CDC using RAA RR for the performance parameter Overall Response Time (OvrallResTime) and Data Centre Process Time (DCPT), respectively (Figs. 4 and 5).

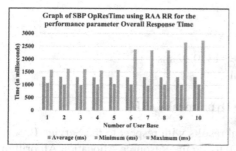

Fig. 4. Graph of SBP OptiResTime using RAA RR for the performance parameter Overall Response Time (OvrallResTime).

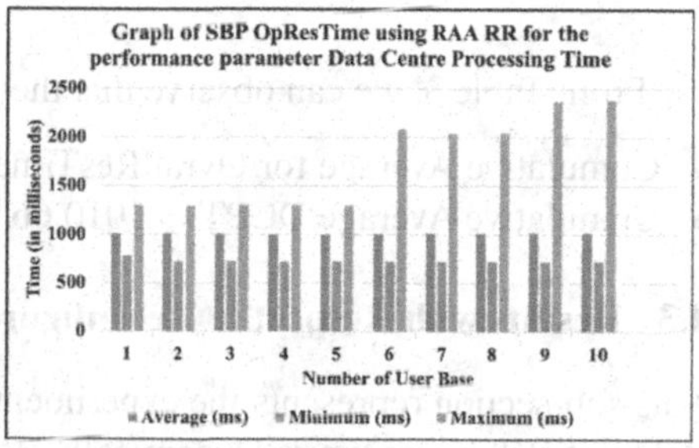

Fig. 5. Graph of SBP OptiResTime using RAA RR for the performance parameter Data Centre Processing Time (DCPT)

Table 1 represents the experimental results concerning SBP OptiResTime using the RAA RR (Table 3).

Table 3. Experimental Results with respect to OptiResTime.

Scenario No.	Overall Response Time (OvrallResTime) (in milliseconds)			Data Centre Processing Time (DCPT) (in milliseconds)		
	Average (ms)	Minimum (ms)	Maximum (ms)	Average (ms)	Minimum (ms)	Maximum (ms)
1	1311.41	1068.01	1583.52	1009.89	780.01	1300.01
2	1312.4	1005.01	1622.53	1012.21	720.51	1300.52
3	1306.1	988.51	1610.02	1005.27	720.51	1300.51
4	1301.36	1021.51	1568.52	1000.11	720.01	1300.01
5	1301.29	1041.01	1586.02	1001.49	720.51	1300.51
6	1307.91	1018.51	2387.53	1008.24	720.01	2080.02

(continued)

Table 3. (*continued*)

Scenario No.	Overall Response Time (OvrallResTime) (in milliseconds)			Data Centre Processing Time (DCPT) (in milliseconds)		
	Average (ms)	Minimum (ms)	Maximum (ms)	Average (ms)	Minimum (ms)	Maximum (ms)
7	1313.3	976.51	2358.03	1013.47	720.01	2040.01
8	1314.57	1021.51	2361.56	1014.84	720.01	2045.05
9	1319.61	1014.01	2666.56	1019.4	720.01	2365.03
10	1321.2	1032.01	2747.59	1021.72	720.51	2380.58

From Table 3, we can observe that the:

- Cumulative Average for OvrallResTime = 1310.92 ms
- Cumulative Average DCPT = 1010.66 ms

4.3 Results with Respect to Reconfigure Dynamically (RD)

This sub-section represents the experimental results with respect to the Service Broker Policy (SBP) Reconfigure Dynamically (RD) using the Resource Allocation Algorithm (RAA) Round – Robin (RR). Figure 2 and 3 represents the graph of SBP CDC using RAA RR for the performance parameter Overall Response Time (OvrallResTime) and Data Centre Process Time (DCPT), respectively (Figs. 6 and 7).

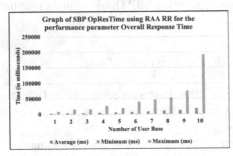
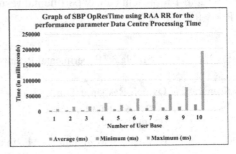

Fig. 6. Graph of SBP RD using RAA RR for the performance parameter Overall Response Time (OvrallResTime).

Fig. 7. Graph of SBP RD using RAA RR for the performance parameter Data Centre Processing Time (DCPT).

Table 4 represents the experimental results concerning SBP Reconfigure Dynamically (RD) using the RAA RR.

From Table 4, we can observe that the

- Cumulative Average for OvrallResTime = 8226.72 ms
- Cumulative Average DCPT = 7925.28 ms

Table 4. Experimental Results with respect to RD

Scenario No.	Overall Response Time (OvrallResTime) (in milliseconds)			Data Centre Processing Time (DCPT) (in milliseconds)		
	Average (ms)	Minimum (ms)	Maximum (ms)	Average (ms)	Minimum (ms)	Maximum (ms)
1	4413.34	1210.52	8984.51	4113.65	900.51	8680.5
2	4871.35	1158.02	15934.02	4571.85	840.01	15610.51
3	4774.28	1123.52	15922.02	4473.22	840.51	15626.51
4	5348.81	1123.52	26824.52	5047.14	840.51	26535.51
5	6223.95	1141.51	19477.52	5920.5	840.51	19161.01
6	6875.72	1138.52	40498.52	6571.56	840.02	40218.51
7	8505.26	1117.52	47101.52	8202.47	840.51	46781.01
8	10134.92	1143.02	53098.52	9834.07	850.51	52819.51
9	12463.54	1135.52	75009.58	12161.09	850.01	74710.07
10	18655.99	1133.52	191431.26	18357.26	840.01	191125.25

5 Cumulative Results and Reinforcement Learning to Improve Cloud Performance

This section includes the cumulative results of the entire experiment along with suggesting the Reinforcement Learning (RL) method to enhance the SBPs. Table 5 represents the final results of the experiment.

Table 5. Final results of the experiment.

Parameter	Overall Response Time (OvrallResTime)			Data Center Processing Time (DCPT)		
	Average (ms)	Minimum (ms)	Maximum (ms)	Average (ms)	Minimum (ms)	Maximum (ms)
CDC	1310.68	1023.32	2085.04	1010.71	726.46	1777.73
OptiResTime	1310.92	1018.66	2049.19	1010.66	726.21	1741.23
RD	8226.72	1142.52	49428.2	7925.28	848.31	49126.84
Final Results	CDC > OptiResTime > RD	OptiResTime > CDC > RD	OptiResTime > CDC > RD	OptiResTime > CDC > RD	OptiResTime > CDC > RD	OptiResTime > CDC > RD

From the above table, we can observe that the performance of Closest Data Center (CDC) is better than that of OptiResTime, followed by the performance of RD concerning the performance parameter Overall Response Time (OvrallResTime). Concerning the second performance parameter, the performance of OptiResTime is better than the Closest Data Center, followed by the performance of RD. The RD offers the most minor performance in both the performance parameters, making it a less acceptable approach. In this current system, the SBPs are statically fixed and remain the same for processing all the user tasks. These SBPs do not offer any intelligence mechanism to enhance the cloud's performance. To solve these issues, the Machine Learning (ML) model of Reinforcement Learning (RL) can provide a completely dynamic selection-based method where the SBP will be dynamically selected according to the current cloud's situation. Also, the RL method has been iconic and helpful in ensuring all the faulty tasks can be handled and computed, which is absent in the current system.

Figure 8 depicts the flow diagram for the suggested method using the RL method.

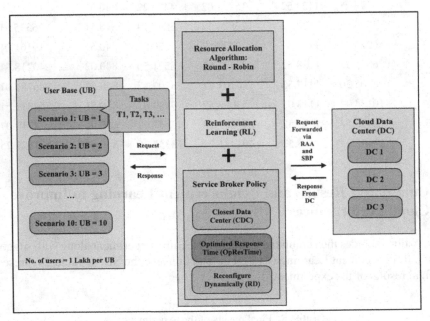

Fig. 8. Flow diagram for the suggested method.

6   Conclusion

In this paper, the researchers have studied and compared the cloud's service broker policies (SBPs) using the resource allocation algorithm Round–Robin (RR). The SBPs considered for this study and comparison are Closest Data Center (CDC), Optimized Response Time (OptiResTime), and Reconfigured Dynamically (RD). The performance parameters that were considered for this study were Overall Response Time (OvrallResTime) and Data Center Processing Time. For this study and comparison, an experiment was conducted in the CloudAnalyst cloud simulation platform where users from their user bases submitted their requests as tasks to the cloud Data Center (DC). These DCs processed and computed these tasks using the SBPs CDC, OptiResTime, and RD. To have a thorough comparison among the used SBPs, this experiment was conducted in ten scenarios, where the number of user bases differed in each scenario. The experimental results conveyed that SBP CDC outperformed others regarding the performance parameter OvrallResTime. Whereas the SBP OpRestTime outperformed other SBPs in terms of performance parameter DCPT. Each SBP possesses some positives and negatives concerning each performance parameter. Hence, the Machine Learning (ML) model of Reinforcement Learning (RL) has been suggested to dynamically choose an SBP depending upon the current state of the cloud computing environment. Also, since the RL method is beneficial in unknown situations, it can handle erroneous situations in the cloud computing environment.

References

1. Alwada'n, T., Al-Tamimi, A., Mohammad, A.H., Salem, M., Muhammad, Y.: Dynamic congestion management system for cloud service broker. Int. J. Power Electron. Drive Syst. **13**(1), 872 (2023)
2. Chowdhury, S., Katangur, A.K.: Optimization of Datacenter Selection Policy in Cloud Computing using Differential Evolution Algorithm. IEEE (2023)
3. Goel, G., Chaturvedi, A.K.: A Comprehensive Review of QoS Aware Load Balancing Techniques in Generic & Specific Fog Deployment Scenarios. IEEE (2023)
4. Nayak, B., Bisoyi, B., Pattnaik, P.K.: Data center selection through service broker policy in cloud computing environment. Mater. Today Proc. **80**, 2218–2223 (2023)
5. Shahid, M.A., Alam, M.M., Su'ud, M.B.M.: Performance evaluation of load-balancing algorithms with different service broker policies for cloud computing. Appl. Sci. **13**(3), 1586 (2023)
6. Zhang, H., Sun, R.: A novel optimal management method for smart grids incorporating cloud-fog layer and honeybee mating optimization algorithm. Sol. Energy **262**, 111874 (2023)
7. Elrotub, M., Gherbi, A.: Multi-cloud service brokers for selecting the optimal data center in cloud environment. Int. J. Cloud Appl. Comput. **12**(1), 1–19 (2022)
8. Panda, S.K., Ramesh, K., Indraneel, K., Ramu, M., Damayanthi, N.N.: Novel Service Broker and Load Balancing Policies for CloudSim-Based Visual Modeller. IEEE (2022)
9. Parida, S., Pati, B., Nayak, S.C., Panigrahi, C.R.: EMRA: an efficient multi-optimization-based resource allocation technique for infrastructure cloud. J. Ambient. Intell. Humaniz. Comput. **14**(7), 8315–8333 (2022)
10. Sanjalawe, Y.K., Anbar, M., Al-Emari, S., Abdullah, R., Hasbullah, I.H., Aladaileh, M.: Cloud data center selection using a modified differential evolution. Comput. Mater. Continua **69**(3), 3179–3204 (2021)
11. Jiang, Y., Kodialam, M., Lakshman, T.V., Mukherjee, S., Tassiulas, L.: Resource allocation in data centers using fast reinforcement learning algorithms. IEEE Trans. Netw. Serv. Manage. **18**(4), 4576–4588 (2021)
12. Li, J., Zhang, X., Wei, Z., Wei, J., Ji, Z.: Energy-aware task scheduling optimization with deep reinforcement learning for large-scale heterogeneous systems. CCF Trans. High Perform. Comput. **3**(4), 383–392 (2021)
13. Barhate, S.M., Dhore, M.P.: Hybrid Cloud: A Cost Optimised Solution to Cloud Interoperability. IEEE (2020)
14. Jyoti, A., Shrimali, M.D., Tiwari, S., Singh, H.: Cloud computing using load balancing and service broker policy for IT service: a taxonomy and survey. J. Ambient. Intell. Humaniz. Comput. **11**(11), 4785–4814 (2020)
15. Khodar, A., Mager, V.E., Alkhayat, I., Al-Soudani, F.A.J., Desyatirikova, E.N.: Evaluation and Analysis of Service Broker Algorithms in Cloud-Analyst. IEEE (2020)

16. Parida, S., Pati, B.: A cost-efficient service broker policy for data center allocation in IAAS cloud model. Wireless Pers. Commun. **115**(1), 267–289 (2020)
17. Sanaj, M.S., Prathap, P.M.J.: An Enhanced Round Robin (ERR) algorithm for Effective and Efficient Task Scheduling in cloud environment. IEEE (2020)
18. Guroob, A.H., Shetty, A., Manjaiah, D.H.: A research on the effectiveness of the different algorithms and the scheduling in improving the performance of cloud computing by using cloud analyst simulator. Int. J. Innov. Technol. Explor. Eng. **8**(6S4), 1170–1176 (2019)
19. Belgaum, M.R., Musa, S., Su'ud, M.M., Alam, M.M.: A behavioral study of task scheduling algorithms in cloud computing. Int. J. Adv. Comput. Sci. Appl. **10**(7) (2019)
20. Benlalia, Z., Abderahim, B., Karim, A., Ezzati, A.: A new service broker algorithm optimizing the cost and response time for cloud computing. Procedia Comput. Sci. **151**, 992–997 (2019)
21. Desyatirikova, E.N., Khodar, A., Rechinskiy, A.V., Chernenkaya, L.V., Alkhayat, I.: Performance analysis of available service broker algorithms in Cloud Analyst. In: Arseniev, D.G., Overmeyer, L., Kälviäinen, H., Katalinić, B. (eds.) CPS&C 2019. LNNS, vol. 95, pp. 449–457. Springer, Cham (2020). https://doi.org/10.1007/978-3-030-34983-7_44
22. Jyoti, A., Shrimali, M.D.: Dynamic provisioning of resources based on load balancing and service broker policy in cloud computing. Clust. Comput. **23**(1), 377–395 (2019)
23. Kofahi, N.A., Alsmadi, T., Barhoush, M., Al-Shannaq, M.A.: Priority-based and optimized data center selection in cloud computing. Arab. J. Sci. Eng. **44**(11), 9275–9290 (2019)
24. Pandey, A., Lyu, Z., Joshi, T., Calyam, P.: OnTimeURB: Multi-Cloud Resource Brokering for Bioinformatics Workflows. IEEE (2019)
25. Pj, K., Sivakumar, N., Prabhu, J., Ramesh, P.: Analysis of heterogeneous device characteristics in round robin based load balancing algorithm with closest data center as service broker policy in cloud. Int. J. Innov. Technol. Explor. Eng. **8**(9), 1627–1630 (2019)
26. Kumar, P.J., Suganya, P., Malhotra, K., Yadav, P.: A suite of load balancing algorithms and service broker policies for cloud: a quantitative analysis with different user grouping factor in cloud. Int. J. Recent Technol. Eng. **8**(1), 2983–2989 (2019)
27. Anuragi, R., Pandey, M.K.: Review paper on cloudlet allocation policy. In: Advances in Intelligent Systems and Computing, pp. 319–327 (2018)
28. Ashraf, M.H., Javaid, N., Abbasi, S.H., Rehman, M., Sharif, M.U., Saeed, F.: Smart grid management using cloud and fog computing. In: Barolli, L., Kryvinska, N., Enokido, T., Takizawa, M. (eds.) NBiS 2018. LNDECT, vol. 22, pp. 624–636. Springer, Cham (2019). https://doi.org/10.1007/978-3-319-98530-5_54
29. Fatima, A., Javaid, N., Waheed, M., Nazar, T., Shabbir, S., Sultana, T.: Efficient resource allocation model for residential buildings in smart grid using FOG and cloud computing. In: Advances in Intelligent Systems and Computing, pp. 289–298 (2018)
30. Kapoor, L., Jindal, A., Benslimane, A., Aujla, G.S., Chaudhary, R., Kumar, N., Zomaya, A.Y.: SLOPE: a self learning optimization and prediction ensembler for task scheduling. Networking Commun. (2018)
31. Meftah, A., Youssef, A.E., Zakariah, M.: Effect of service broker policies and load balancing algorithms on the performance of large-scale internet applications in cloud datacenters. Int. J. Adv. Comput. Sci. Appl. **9**(5) (2018)
32. Naeem, M., Javaid, N., Zahid, M., Abbas, A., Rasheed, S., Rehman, S.: Cloud and fog based smart grid environment for efficient energy management. In: Xhafa, F., Barolli, L., Greguš, M. (eds.) INCoS 2018. LNDECT, vol. 23, pp. 514–525. Springer, Cham (2019). https://doi.org/10.1007/978-3-319-98557-2_48
33. Patel, R., Patel, S.: Efficient service broker policy for intra datacenter load balancing. In: Smart Innovation, Systems and Technologies, pp. 683–692 (2018)

34. Rehman, M., Javaid, N., Ali, M.J., Saif, T., Ashraf, M.H., Abbasi, S.H.: Threshold based load balancer for efficient resource utilization of smart grid using cloud computing. In: Xhafa, F., Leu, F.Y., Ficco, M., Yang, C.T. (eds.) 3PGCIC 2018, pp. 167–179. Springer, Cham (2018). https://doi.org/10.1007/978-3-030-02607-3_16

35. Rekha, P., Dakshayini, M.: Dynamic cost-load aware service broker load balancing in virtualization environment. Procedia Comput. Sci. **132**, 744–751 (2018)

36. Manasrah, A.M., Aldomi, A., Gupta, B.B.: An optimized service broker routing policy based on differential evolution algorithm in fog/cloud environment. Clust. Comput. **22**(S1), 1639–1653 (2017)

37. Armbrust, M., et al.: A view of cloud computing. Commun. ACM **53**(4), 50–58 (2010)

38. Dillon, T.S., Wu, C., Chang, E.: Cloud Computing: Issues and Challenges. IEEE (2010)

39. Vengerov, D.: A reinforcement learning approach to dynamic resource allocation. Eng. Appl. Artif. Intell. **20**(3), 383–390 (2007)

40. Andrew, A.M.: Reinforcement learning. Kybernetes **27**(9), 1093–1096 (1998)

A Course Study Planning Framework – An Engineering Perspective

Mohammad Shakeel Laghari[✉], Addy Wahyudie, Ahmed Hassan,
Abdulrahman Alraeesi, and Mahmoud Haggag

College of Engineering, United Arab Emirates University, Abu Dhabi, UAE
mslaghari@uaeu.ac.ae

Abstract. Understudy course enrollment is a significant as well as a paltry cycle to guarantee a study satisfies the degree prerequisites of a college in a complete and organized manner without experiencing superfluous postponements. Most academic institutions burn through a significant measure of cash for customized exhorting and enrollment projects to suit their requirements. A stronger course counseling and scheduling structure could help engineering students at the United Arab Emirates University's College of Engineering create effective review schedules. Albeit the school has a modified enlistment framework called Banner and a distinct course arranging framework called DegreeWorks, course prompting and arranging at the college and student level frequently miss the mark concerning wanted capabilities, abilities, and capacities. The shortfall of a viable informing framework and the need for suitable direction can present difficulties for students. This exploration has constructed a Course Study Planning Framework (CSPF), a software package to assist students with planning suitable review plans. The framework is intended to permit understudies to pick the most reasonable courses for the accompanying semesters. The consequence of this cycle is that the finished review plans are saved in a record.

Keywords: Academic Course Advising · Course Planning and Scheduling · Study Plans

1 Introduction

Scholastic exhorting and course arranging are supposed to exchange essential data to assist partners with arriving at their instructive and scholarly objectives. It is a comprehension and divided liability between a scholarly counsel and the student. Advice is crucial when a faculty member (staff member) inquires about a college student's academic concerns, such as prerequisites for degrees, course and career planning, and common discourse regarding whether an area of study meets a specific intellectual or vocational interest [1, 2].

Three main elements make up a student's course enrollment process: information on the unavoidable and essential pre-enrollment facts, requirements, and instructions for registering for courses, and lastly, instructions on what to do next once the student has completed the course enrollment process.

© The Author(s), under exclusive license to Springer Nature Switzerland AG 2024
K. K. Patel et al. (Eds.): icSoftComp 2023, CCIS 2031, pp. 214–228, 2024.
https://doi.org/10.1007/978-3-031-53728-8_17

The process of enrolling in classes and the guidelines that alumni must follow are usual; yet, deciding which classes to take is a big decision that requires careful consideration. An understudy often chooses mediocre courses randomly from an infinite supply of useful courses.

Students who take significantly more courses failed on average more often than predicted or changed their course selections for different semesters are included in the normal enrollment questions that are based on course preparation and warning.

Most courses require a heavy burden in addition to the unsuccessful combination of characteristic and course grades, considered, or chosen courses in addition to opportunity conflicts, or even better periods apart, such as early morning and late afternoon classes. Less than normal course selection permits an increase in completion duration, and too many courses permit a heavy burden in addition to the hardship of characteristics and course grades.

Course warnings included in study plans are somewhat addressed before enrollment, but students still experience opportunity conflicts during registration, which commonly leads to failing different options as classes pile up quickly before registration begins. It is also possible for a student to consent to the prospect of completing all coursework on two days of the event's or entity's temporal duration, or alternatively to open smoothly on all four days of the week.

Students who are accompanied by such inquiries may experience a delay in starting their industrial training (individual term-long learning performed in the industry) because of unnecessary course selection or bearing missed assignments in their major courses, losing an entire semester by failing to complete the necessary requirements of a minimum of 12 credits or carrying a heavy semester load operating somewhat because of taking a disproportionate number of courses.

The CSPF was created to provide graduates with precise time-based plans, more options for enrolling in courses they are interested in, and advice on how to satisfy their fitness criteria as quickly as possible. The front-end setup of the operating system is completed by utilizing CSS, HTML, and JavaScript. For one Node.js foundation and one Express.js atheneum, server-side setup is implemented. The NoSQL Database System (MongoDB) is utilized to maintain study plans.

The developed software helps students select the right courses (ranging from four to seven) to build a study program. The first stage in the typical course selection procedure is for the student to upload a list of the courses they have finished and passed in the current semester. After that, the program goes through several steps to provide pupils with a nearly perfect list of the top classes. The courses in the curriculum are organized around a body of information that directs course selection.

The article describes the functioning of the whole course scheduling and planning package, including priority selection of fitting courses, help menus, limits, and all that. In testing, the software package suite generated fifteen practically ideal study schedules. Two such ideas are discussed in this paper as examples.

2 Literature Review

During earlier registration processes, the academic departments of the school controlled enrolling students in courses. Those early operations were overwhelmed by delays, an abundance of resources, and management and administrative issues. Subsequently, numerous viewpoints were used to analyze this insight of registration, including those that were relevant to student course advising, course planning, class scheduling, course registration, etc.

All the aforementioned problems doubled in size in the late 1990s as more students began to enroll. To maximize the distribution of course spaces for this increased number of registered students, the registration process also required reworking because of this increment. Additionally, a lot of the universities began to provide multidisciplinary, specialized, and advanced degree programs. Time conflicts have also been worse when choosing courses for these degrees. Choosing the best courses without running into scheduling problems is the major goal of course planning and registration systems.

Computerized registration techniques were developed and gained acceptance over time. Even though these systems provide many benefits, the main disadvantage was having to deal with frequent machine and network outages. However, it had been expected that some "offline" data processing would be necessary for faults that had not been anticipated. Most human errors, such as those involving the accurate entry of data, were anticipated to be identified by the system as they happened. The following paragraphs examine a few of the early systems.

Given that creating a course plan is a challenging optimization problem, genetic algorithms are employed to build an information strategy for course scheduling. By examining potential alternatives, an ideal solution is sought. The "*Codeigniter*" framework, Responsive Bootstrapping for the user interface, and MySQL for the database make up the genetic algorithm method's primary configuration. This technique helps to greatly eliminate course conflicts [3].

The authors of this reference paper researched many academic systems in order to create an automated system for academic advising in a typical educational system. An impression of the development and use of a new model is discussed as a web-based application. To help advisors and other advising staff respond to student complaints and ideas, a model is offered. The model, a computerized system, enables academic advisors to provide their pupils with exact, superior guidance [4].

To create multi-year course plans for complicated situations like study-abroad semesters, double majors, transfer credits, course overrides, and early graduation, a TAROT system for course advising was created. The design and functionality of TAROT are discussed in this essay, along with how it differs from other course advising systems in terms of its features [5].

The main authors of the current study developed, implemented, and evaluated several course planning and advising systems beginning in 2005. Using criteria that varied in the number of priority fields each course had as well as in the arrangement of these fields, algorithms for course design and advising were studied. The majority of these systems have been phased out as a result of curricular changes and a few unanticipated errors [6–10].

This paper proposes a recommender system for advisors and students to develop personalized study programs for each student over several semesters. The proposed remedy includes ML-related ideas, explainable suggestions, performance modeling, graph theory, and an intuitive user interface [11].

In this paper, a framework for booking courses assisted by artificial awareness is understood by a few designers. This study also discusses the shortcomings of the current planning framework and compares them to the new plan. The course planning framework paradigm is implemented using a program/server mode approach [12].

The researchers in this reference paper created a student information system called MYGJU that can manage both academic and financial data. The user-friendly online course registration process prevents students from choosing inappropriate or superfluous courses. Even though MYGJU was introduced, the system did not become popular among students, as is also covered in this reference work [13].

The writers of this review examine, select, and modify effective course-exhorting improvements to present a clever personalized course prompting model. By incorporating standard plans for all higher education organizations and making the standard, guidelines, educational program, and reviewing framework more adaptable to modify without impacting unrelenting scholarly information, it defeats the current course prompting frameworks. The suggested strategy helps students by categorizing and selecting appropriate courses based on their academic backgrounds and individual interests [14].

For students wishing to apply for a university degree, the writers of this citation paper have suggested designing an online course selection system. They also suggest adding enhancements to a rule-based decision-making system based on this design. When deciding between a public and a private university, this design permitted them to make an arguably better choice [15].

The authors of this reference paper researched several academic systems in order to create an automated system for academic advising in a typical university system. A firsthand account of the development and execution of a new model is addressed as a web-based application. A template has been provided to make it simpler for advisors and other advertising staff to respond to complaints and ideas from students. The model is a computerized framework that helps academic advisers provide reliable, accurate, and high-quality counseling to their students [4].

3   Course Level Priorities

The United Arab Emirates University (UAEU) has about 15,000 students enrolled, whereas the College of Engineering (COE) can hold about 3,000 students. The UAEU is made up of nine colleges, including the COE. Electrical and Communication Engineering is one of the five COE departments that has roughly 400 permanent students.

The students finish 147 credits of coursework to meet the requirements for their bachelor's degrees. Contingent upon the type of grades and GPA they receive, it normally takes them 4½ years with an average of 17 credit hours of workload per semester, 5 years with an average of 16 credits of coursework, 5½ years with an average of 15 course credits, and 6 years with an affirm age of 14 credits per semester to complete the prerequisites for their certificate. Poor academic performers may need longer because they often fail courses.

Students can choose from 52 courses divided into 6 categories: Seven general education courses totaling 21 credits, 15 college requirements accumulating 38 credits, 23 compulsory department courses totaling 55 credits, seven laboratory courses, one industrial training course of 15 credits, two graduation projects totaling 6 credits, and four elective specialization courses totaling 12 credits make up the requirements for a bachelor's degree.

Figure 1 shows the hierarchy charts for all four EE disciplines of computer, electronic, power and control, and communication engineering, respectively. Both required and elective courses are listed in the chart. The background courses for the college are colored green, the required courses and labs are blue, and the technical electives are orange. The arrows denote hierarchies. The courses that are marked on the course box with the symbol " ✦" as being required for industrial training. The alphabet "F" designates the courses offered in the Fall semester and "S" for the Spring semester.

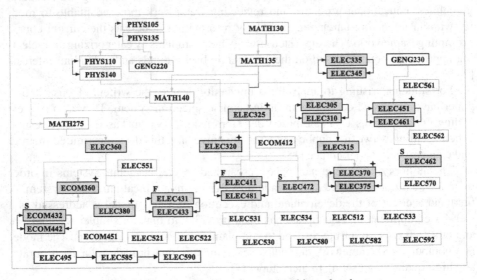

Fig. 1. Electrical Engineering courses hierarchy chart.

Figure 2 shows the software package's user interface. The Electrical Engineering disciplines (groups) of Control, Power, Electronics, Communication, and Computer Engineering are divided into different courses. College and general education courses are on the left side of the GUI, and major courses are in the center. The start semester for constructing the study plans is specified in one section. The right window shows the results of the selection.

A student enrolled in an Industrial Training (IT) course works with an industrial unit during the whole semester, including the summer. Depending on one's academic status and after completing 93 credits or more, one becomes qualified for the hands-on industrial training phase.

In order to advance the knowledge field of course design, each course has three priority levels allocated to it, as shown in Table 1 and Table 2, respectively. When picking courses for the development of study plans, these criteria help to define individual courses in terms of their importance. These are arranged from highest to lowest priority in the following order:

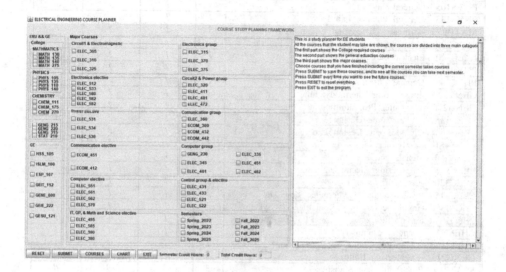

Fig. 2. Interface of the software package.

3.1 Priority of Industrial Training

Industry-related Training (IT) is connected to the first level of importance. During the course selection process, this quality is given the highest emphasis. Delaying an IT phase could put off graduating. All courses that must be completed before IT is represented by the number "1" as shown in Table 1. For instance, IT requires ELEC 320 to be completed but not ELEC 472. The courses in Table 2 are not necessary, whereas the courses in Table 1 are all required for industrial training.

3.2 Priority of Hierarchy

This relates to the hierarchy of course levels for each course. For instance, according to Fig. 1 and Table 1, MATH135 comprises four hierarchical layers. The hierarchies are displayed as:

- MATH135 -> (MATH140/MATH275) -> ELEC360 -> ELEC380,
- MATH135 -> (MATH140/MATH275) -> ELEC360 -> (ECOM360/ECOM432),
- MATH135 -> ELEC325.

Students who miss classes at high hierarchical levels may put off their graduation and professional development.

Table 1. All courses that are required for IT with three priority levels.

Course Code	CH	Required for IT	Levels of hierarchy	Courses to open
MATH130 – Calculus_I_for_Engineering	3	1	5	22
MATH135 – Calculus_II_for_Engineering	3	1	4	19
MATH140 – Linear_Algebra_I	3	1	3	8
PHYS105 – General_Physics_I	3	1	3	8
PHYS135 – General_Physics_Lab_I	1	1	3	8
MATH275 – Ordinary_Differential_Equations	3	1	2	7
ELEC305 – Electric_Circuits_I	3	1	2	7
ELEC310 – Electric_Circuits_I_Lab	1	1	2	7
PHYS110 – General_Physics_II	3	1	2	4
PHYS140 – General_Physics_Lab_II	1	1	2	4
GENG220 – Engineering_Thermodynamics	3	1	2	4
GENG230 – Computer_Programming	3	1	2	3
ELEC335 – Digital_Logic_Design	3	1	2	3
ELEC345 – Digital_Logic_Design_Lab	1	1	2	3
ELEC451 – Microprocessors	3	1	2	2
ELEC360 – Signals_and_Systems	3	1	1	5
ELEC325 – Engineering_Electromagnetics	3	1	1	3
ELEC320 – Electric_Circuits_II	3	1	1	3
ELEC315 – Fundamentals_of_Microelectronic_Devices	3	1	1	2
ECOM360 – Fundamentals_of_Communication_Systems	3	1	1	2
ESPU107 – Introduction_of_Academic_English_for_Engg.	3	1	1	1
CHEM111 – General_Chemistry_I	3	1	1	1
CHEM175 – Chemistry_Lab_I_for_Engineering	1	1	1	1
GENG215 – Engineering_Ethics	2	1	0	0
STAT210 – Probability_and_Statistics	3	1	0	0
GENG315 – Engineering_Economics	3	1	0	0
CHEM270 – Materials_Science	3	1	0	0
ELEC370 – Electronic_Circuits	3	1	0	0
ELEC375 – Electronic_Circuits_Lab	1	1	0	0
ELEC380 – Analytical Methods for Electrical Engineering	3	1	0	0

3.3 Priority of Open Courses

This has to do with the total number of courses that are linked to a specific course. A numerical value of "1" is assigned to each course that a particular course opens. MATH140 oversees the opening of seven different courses at all linked hierarchical levels, as shown in Fig. 1. These courses are ELEC360, ECOM360, ELEC380, ECOM432, ECOM442, ELEC431, and ELEC433. The total number count includes the associated lab courses as well. A high number here indicates how crucial a course is to open with that many courses.

Table 2. All courses that are not required for IT with three priority levels.

Course Code	CH	Required for IT	Levels of hierarchy	Courses to open
ELEC461 – Microprocessors_Lab	1	0	1	1
ELEC585 – Design_&_Critical_Thinking_in_Elec._Engg.	3	0	1	1
ECOM432 – Data_Communications_&_Networks	3	0	0	0
ECOM442 – Data_Communications_&_Networks_Lab	1	0	0	0
ELEC411 – Electric_Energy_Conversion	3	0	0	0
ELEC431 – Control_Systems	3	0	0	0
ELEC433 – Instrumentation_and_Control_Lab	1	0	0	0
ELEC462 – Computer_Architecture_and_Organization	3	0	0	0
ELEC472 – Power_Systems	3	0	0	0
ELEC481 – Electric_Energy_Conversion_Lab	1	0	0	0
HSS105 – Emirates_Studies	3	0	0	0
ISLM100 – Islamic_Culture	3	0	0	0
GESU121 – Sustainability	3	0	0	0
GEIE222 – Funda._of_Innovation_and_Entrepreneurship	3	0	0	0
GEIT112 – Fourth_Industrial_Revolution	3	0	0	0
GENE000 – General_Education (Hum. & Fine_Arts_Area)	3	0	0	0
ELEC495 – Industrial_Training	15	0	0	0
ELEC590 – Capstone_Engineering_Design_Project	3	0	0	0

4   The CSPF Package

The usual student chooses all finished and ongoing semester courses when they first launch the xx software procedure. The information in the database also includes the student's full name, institution ID, email address, year of readiness for the study plan, and the name of the college in conjunction with the course selection.

The database system initially creates three distinct sets of files. The curriculum's courses are included in the first file along with their course IDs, titles, credit hours, and precedence levels. The courses available during the Spring semester are in a third file, whereas those given just during the Fall semester are in a second file.

The framework, as shown in Fig. 3, has several clickable buttons. Two of them, titled "Course" and "Chart," when clicked, respectively, disclose the courses listed in Table 1 and the hierarchical charts. By selecting the checkboxes, students can select their courses utilizing the data structures.

To choose the starting semester for the study plan, a different checkbox menu is used. The number of credits in the currently chosen semester and the overall number of credits are displayed in two text fields at the center bottom package image. By deselecting the course button, a mistake in course selection can be corrected, and the course credits are then removed from the semester credit hours. The 'RESET' button, however, allows the entire process to be repeated if a mistake is found after pressing the 'SUBMIT' button.

The package also recommends top-priority courses for the upcoming semesters. This recommendation is based on the idea that the student should only be shown required courses for which they have met all necessary prerequisites. By only showing courses that are acceptable for a certain semester, this recommendation aids the student in avoiding selecting courses that are not available for that semester. Additionally, the list of suggestions is ranked from most important course to least important course, and so on.

In Tables 3, 4, and 5 three test examples of study plans are presented.

4.1 Example Plan 1

Table 3 (a) lists the courses that have been completed and are presently being taken. To distinguish between Requirements, the courses are colored differently. The study schedule starts in the Spring of 2023 and runs through the Fall of 2025. The three priority levels are used to determine which courses to take. First-level check lays out that the course is vital for vocational preparation and that deferring it could postpone graduation.

Course hierarchy is the second priority. A greater number connected with a particular course indicates that the course has numerous levels of forward hierarchy. Opening many courses that are dependent on the chosen one is the third priority. In this case, the importance of the course to be taken earlier is demonstrated by a bigger number associated with a course, as can be observed in Tables 1 and 2.

The first selection process is displayed in Fig. 6. The checkboxes for the current semester and the completed courses are checked. Additionally, the checkbox for the Spring 2023 beginning semester is selected. The selection results are displayed using the SUBMIT button and show that 33 credits have already been earned.

There are also suggestions for the spring 2023 semester. The first seven courses on the list, as given in Table 3 (b), are automatically chosen by the student. The selection process for succeeding semesters can be repeated until the student has earned 147 credits (Table 3 (h)).

This student, who began his bachelor's degree program in the Spring of 2022, anticipates completing it in seven semesters, including the summer of industrial training if the right number of credits for each successive semester is chosen.

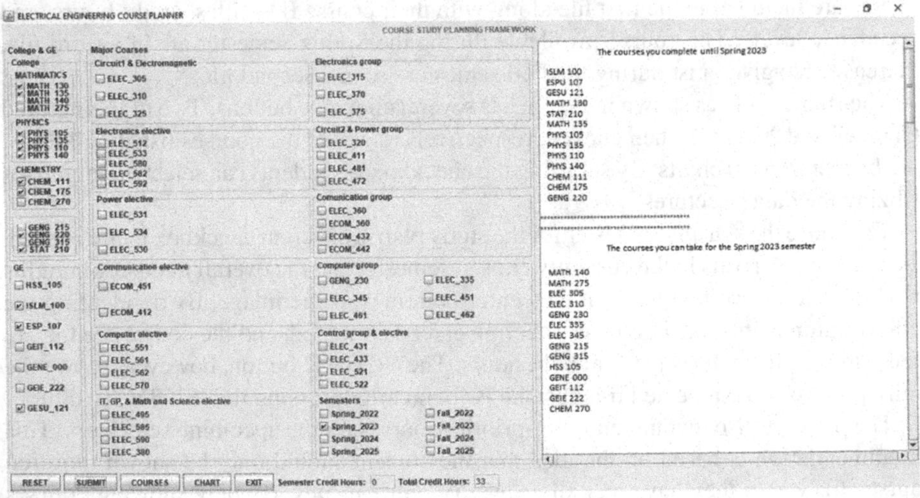

Fig. 3. The selection process to create study plans.

Table 3. Study schedule for Example Plan 1.

(a) Passed and current semester courses:

#	Course Code	CH	#	Course Code	CH
1	ISLM100	3	2	GESU121	3
3	ESPU107	3	4	MATH130	3
5	PHYS105	3	6	PHYS135	1
7	CHEM111	3	8	CHEM175	1
9	PHYS110	3	10	PHYS140	1
11	STAT210	3	12	MATH135	3
13	GENG220	3			

Total Credits: 33

(b) Spring 2023 Semester:

#	Course Code	CH	#	Course Code	CH
1	MATH140	3	2	MATH275	3
3	ELEC305	3	4	ELEC310	1
5	ELEC335	3	6	ELEC345	1
7	GENG230	3			

Semester Credits: 17   Total Credits: 50

(c) Fall 2023 Semester:

#	Course Code	CH	#	Course Code	CH
1	ELEC315	3	2	ELEC320	3
3	ELEC325	3	4	ELEC360	3
5	ELEC451	3	6	GENG215	2

Semester Credits: 17   Total Credits: 67

(d) Spring 2024 Semester:

#	Course Code	CH	#	Course Code	CH
1	ELEC461	1	2	ECOM360	3
3	ELEC370	3	4	ELEC375	1
5	ELEC380	3	6	CHEM270	3
7	GENG315	3			

Semester Credits: 17   Total Credits: 84

(e) Fall 2024 Semester:

#	Course Code	CH	#	Course Code	CH
1	ELEC411	3	2	ELEC481	1
3	ELEC431	3	4	ELEC433	1
5	Elective1	3	6	HSS105	3
7	GENE000	3			

Semester Credits: 17   Total Credits: 101

(f) Spring 2025 Semester:

#	Course Code	CH	#	Course Code	CH
1	ECOM432	3	2	ECOM442	1
3	ELEC462	3	4	ELEC472	3
5	Elective2	3	6	ELEC585	3

Semester Credits: 16   Total Credits: 117

(g) Summer 2025:

#	Course Code	CH	#	Course Code	CH
1	ELEC495	15			

Semester Credits: 15   Total Credits: 132

(h) Fall 2025 Semester:

#	Course Code	CH	#	Course Code	CH
1	GEIE222	3	2	GEIT112	3
3	Elective3	3	4	Elective4	3
5	ELEC590	3			

Semester Credits: 15   Total Credits: 147

4.2 Example Plan 2

This student has a low GPA and is only permitted to enroll in 13 to 16 credits each semester despite having taken 58 credits worth of curriculum. The student will need an additional six semesters (including the summer) to finish the prerequisites for the degree. All semesters have 15 or fewer credits, except the Spring semester of 2024. The complete study plan is shown in Table 4.

4.3 Example Plan 3

The project methodology is explained in depth regarding Example Plan 3. Before enrolling for the Spring 2023 semester, this average student had accrued 43 credits, as shown in Table 5 (a).

Only the chosen courses are shown for each upcoming semester in the first two examples. Here, all of the courses that this understudy is eligible to attend are listed. Even though there are many more courses available, just 13 are appropriate for this particular student because they meet the prerequisites. The courses that are given in Table 5 (b) are in the same order as those that are listed in Tables 1 and 2.

Reviewing MATH140 reveals that it has three levels of hierarchy, opens eight courses at all levels, and is a prerequisite for technical training. This course is the most sought-after to be taken as soon as possible due to these three qualities. The courses following MATH140 are also arranged in a prioritized manner to make it simpler for the student to choose, beginning at the top and continuing until the desired number of credits is achieved.

The student wants to take between five and six courses or roughly 15 to 16 credits for a well-rounded semester. The last four courses (which are not necessary for IT) are automatically removed from the nine courses with "Priority for IT." From the subsequent "levels of hierarchy," there are only seven courses left. The final option, "courses to open," determines which six courses will be chosen initially for the Spring 2023 semester. Table 5 (c) shows this selection as courses that have been checked.

For the subsequent semesters, the course selection process uses a similar mechanism. Although IT should have been chosen for the Spring 2025 semester (Table 5(g)), choosing IT in the summer and conventional courses for this semester shortened the time it took to graduate by one semester.

Table 4. Study schedule for Example Plan 2.

(a) Passed and current semester courses:

#	Course Code	CH	#	Course Code	CH
1	ISLM100	3	2	GESU121	3
3	ESPU107	3	4	MATH130	3
5	PHYS105	3	6	PHYS135	1
7	CHEM111	3	8	CHEM175	1
9	PHYS110	3	10	PHYS140	1
11	GENG215	2	12	MATH135	3
13	GENG220	3	14	MATH140	3
15	HSS105	3	16	GEIE222	3
17	GENG220	3	18	CHEM270	3
19	ELEC305	3	20	ELEC310	1
21	ELEC335	3	22	ELEC345	1
23	GENG230	3			
				Total Credits: 58	

(b) Spring 2023 Semester:

#	Course Code	CH	#	Course Code	CH
1	ELEC315	3	2	MATH275	3
3	ELEC320	3	4	ELEC325	3
5	ELEC451	3			
Semester Credits: 15				Total Credits: 73	

(c) Fall 2023 Semester:

#	Course Code	CH	#	Course Code	CH
1	ELEC360	3	2	ELEC370	3
3	ELEC375	1	4	ELEC380	3
5	ELEC461	1	6	GENG315	3
Semester Credits: 14				Total Credits: 87	

(d) Spring 2024 Semester:

#	Course Code	CH	#	Course Code	CH
1	ECOM360	3	2	ECOM432	3
3	ECOM442	1	4	ELEC462	3
5	ELEC472	3	6	Elective1	3
Semester Credits: 16				Total Credits: 103	

(e) Fall 2024 Semester:

#	Course Code	CH	#	Course Code	CH
1	ELEC431	3	2	ELEC433	1
3	ELEC411	3	4	ELEC481	1
5	Elective2	3	6	ELEC585	3
Semester Credits: 14				Total Credits: 117	

(f) Spring 2025 Semester:

#	Course Code	CH	#	Course Code	CH
1	GEIT112	3	2	Elective3	1
3	GENG000	3	4	Elective4	3
5	ELEC590	3			
Semester Credits: 15				Total Credits: 132	

(g) Summer 2025:

#	Course Code	CH	#	Course Code	CH
1	ELEC495	15			
Semester Credits: 15				Total Credits: 147	

Table 5. Study schedule for Example Plan 3.

(a) Passed and current semester courses:

#	Course Code	CH	#	Course Code	CH
1	ESPU107	3	2	PHYS105	3
3	PHYS135	1	4	CHEM111	3
5	CHEM175	1	6	GEIE222	3
7	PHYS110	3	8	PHYS140	1
9	MATH130	3	10	MATH135	3
11	STAT210	3	12	GENG220	3
13	HSS105	3	14	GENG315	3
15	ELEC305	3	16	ELEC310	1
17	GENG230	3			
				Total Credits: 43	

(b)	Course Code	CH	Required for IT	Levels of hierarchy	Courses to open
1	MATH140	3	1	3	8
2	MATH275	3	1	2	7
3	ELEC335	3	1	2	3
4	ELEC345	1	1	2	3
5	ELEC325	3	1	1	3
6	ELEC320	3	1	1	3
7	ELEC315	3	1	1	2
8	GENG215	2	1	0	0
9	CHEM270	3	1	0	0
10	ISLM100	3	0	0	0
11	GEI112	3	0	0	0
12	GESU121	3	0	0	0
13	GENE000	3	0	0	0

(c) Spring 2023 Semester:

#	Course Code	CH	#	Course Code	CH
1	MATH140	3✓	2	MATH275	3✓
3	ELEC335	3✓	4	ELEC345	1✓
5	ELEC325	3✓	6	ELEC320	3✓
7	ELEC315	3	8	GENG215	2
9	CHEM270	3	10	ISLM100	3
11	GEIT112	3	12	GESU121	3
13	GENE000	3			
Semester Credits: 16				Total Credits: 59	

(d) Fall 2023 Semester:

#	Course Code	CH	#	Course Code	CH
1	ELEC451	3✓	2	ELEC360✓	3
3	ELEC315	3✓	4	GENG215✓	2
5	CHEM270	3✓	6	ELEC380	3
7	ISLM100	3	8	GEIT112	3
9	GESU121	3	10	GENE000	3
Semester Credits: 14				Total Credits: 73	

(continued)

Table 5. (*continued*)

(e) Spring 2024 Semester:

#	Course Code	CH	#	Course Code	CH
1	ECOM360	3✓	2	ELEC370	3✓
3	ELEC375	1✓	4	ELEC380	3✓
5	ELEC461	1✓	6	ECOM432	3✓
7	ECOM442	1✓	8	ISLM100	3
9	GEIT112	3	10	GESU121	3
11	GENE000	3			
Semester Credits: 15				Total Credits: 88	

(f) Fall 2024 Semester:

#	Course Code	CH	#	Course Code	CH
1	ELEC411	3✓	2	ELEC431	3✓
3	ELEC433	1✓	4	ELEC481	1✓
5	ISLM100	3✓	6	GEIT112	3✓
7	GESU121	3	8	GENE000	3
9	Elective1	3	10	Elective2	3
Semester Credits: 14				Total Credits: 102	

(g) Spring 2025 Semester:

#	Course Code	CH	#	Course Code	CH
1	ELEC585✓	3✓	2	ELEC495	15
3	ELEC462✓	3✓	4	ELEC472✓	3✓
5	Elective1✓	3✓	6	Elective2✓	3✓
7	GESU121	3	8	GENE000	3
Semester Credits: 15				Total Credits: 117	

(h) Summer 2025:

#	Course Code	CH	#	Course Code	CH
1	ELEC495	15✓			
Semester Credits: 15				Total Credits: 132	

(i) Fall 2025 Semester:

#	Course Code	CH	#	Course Code	CH
1	GESU121	3✓	2	GENE000	3✓
3	Elective3	3✓	4	Elective4	3✓
5	ELEC590	3✓			
Semester Credits: 15				Total Credits: 147	

5 Conclusion

To make sure that students accomplish their bachelor's degree requirements without experiencing unnecessary delays, a suitable student course scheduling system must be established. A course study planning framework has been built based on the three course-level priorities established around each electrical engineering course. By selecting the four to seven appropriate courses per semester, students can create comprehensive study schedules with the aid of this specifically built software.

References

1. Assiri, A., AL-Ghamdi, A., Brdesee, H.: From traditional to intelligent academic advising: a systematic literature review of e-Academic advising. Int. J. Adv. Comput. Sci. Appl. **11**(4), 507–517 (2020)
2. Iatrellis, O., Kameas, A., Fitsilis, P.: Academic advising systems: a systematic literature review of empirical evidence. Educ. Sci. **7**(4), 90 (2017)
3. Wicaksono, F., Putra, B.: Course scheduling information system using genetic algorithms. Inf. Technol. Eng. J. **6**(1), 35–45 (2021)
4. Afify, E., Nasr, M.: A proposed model for a web-based academic advising system. Int. J. Adv. Netw. Appl. **9**(2), 3345–3361 (2017)
5. Anderson, R., Eckroth, J.: Tarot: a course advising system for the future. J. Comput. Sci. Coll. **34**(3), 108–116 (2019)
6. Laghari, M., Memon, Q., Habib-ur-Rehman: Advising for course registration: a UAE university perspective. In: International Proceedings on Engineering Education, Poland (2005)
7. Laghari, M.: EE course planning software system. J. Softw. **13**(4), 219–231 (2018)
8. Laghari, M.: A priority based course planning system for electrical engineering department. Int. J. Inf. Educ. Technol. **8**(8), 546–552 (2018)
9. Laghari, M., Hassan, A.: A software system for smart course planning. Commun. Comput. Inf. Sci. **1788**, 406–418 (2023)
10. Laghari, M., Hraiz, H., Ghebretatios, S., Alshehhi, A.: Academic course planning software system at EECE department. In: ACM International Conference Proceeding Series, pp. 97–104 (2023)
11. Atalla, S., Daradkeh, M., Gawanmeh, A., Khalil, H., Mansoor, W., Miniaoui, S., Himeur, Y.: An intelligent recommendation system for automating academic advising based on curriculum analysis and performance modeling. Mathematics **11**(5) (2023)
12. Huang, M., Huang, H., Chen, I., Chen, K., Wang, A.: Artificial intelligence aided course scheduling system. J. Phys. Conf. Ser. **1792**, 012063 (2021)
13. Al-hawari, F.: MyGJU student view and its online and preventive registration flow. Int. J. Appl. Eng. Res. **12**(1), 119–133 (2017)
14. Tilahun, L., Sekeroglu, B.: An intelligent and personalized course advising model for higher educational institutes. SN Appl. Sci. **2**(1635) (2020)
15. Ghareb, M., Ahmed, A.: An online course selection system: a proposed system for higher education in Kurdistan region government. Int. J. Sci. Technol. Res. **7**(8), 145–150 (2018)

Three Languages Simulate Polygons and Perform on the Web
Python, Fortran, C, with PHP and 'gnuplot'

Miguel Casquilho[1]([✉])[iD] and Pedro Pacheco[2][iD]

[1] Department of Chemical Engineering, IST, 'Universidade de Lisboa' (University of Lisbon), and CERENA, "Centro de Recursos Naturais e Ambiente" (Centre for Natural Resources and the Environment), Lisbon, Portugal
mcasquilho@tecnico.ulisboa.pt
[2] Department of Computer Science and Engineering, IST, 'Univ. de Lisboa' (Univ. of Lisbon), Lisbon, Portugal
pedropacheco@tecnico.ulisboa.pt

Abstract. On a web page, we display, resorting to the Monte Carlo method, the statistical behavior of the perimeter and area of random polygons inscribed in a circle. We obtain three outcomes: those variables behave surprisingly, mainly the perimeter; three languages run at very different speeds, which is expected, but notorious; and the web page is an example of what can be simply done on the Web. The web page is of free access to a user through the use of a browser, thus without any software installation and necessitating no particular power or matching operating system. This is an illustrative example making it possible to ready adaptation to a number of other problems. It combines the Web and convenient computer languages, to wit: PHP, Python, Fortran, and C, and the 'gnuplot' utility for plotting. The study has as objectives: to observe the said behaviors; to serve as a model to solve other problems; and, more generally, to call attention both to the Web as a medium for scientific computing, and to the adequacy of this in scientific publications. This study recommends computing in a web-based style, in general, which can use executable programs mostly similar to classical computing, these being the inevitable difficulty. This scarcely explored approach is promptly accessed here, and invites the cooperation of academia and industry in our technological era.

Keywords: Web-based computing · random generated polygons · computing in Science and Technology · Engineering education · Statistical applications

Supported by (Portuguese) FCT, see Acknowledgements.

K. K. Patel et al. (Eds.): icSoftComp 2023, CCIS 2031, pp. 229–241, 2024.
https://doi.org/10.1007/978-3-031-53728-8_18

1   Introduction

A web-based program, as we mean it, is a computing resource free to use on a dedicated web page[1], where, consequently, a browser is the only necessity, offering the solution to a given class of mathematical problem, where the user supplies particular data to obtain a corresponding result. In the present work, we contribute with a web-based procedure to observe the statistical behavior in a problem in Geometry, also constituting a template for many other mathematical problems.

When we mention "mathematical problems", we are making a distinction: we have experienced the "classical" engineers benefit professionally from cooperating with their informatic counterparts, and reciprocally.

In the sense mentioned, a web-based solver differs from the numerous applications that a user can install on a computer in that it requires neither installation (updating or uninstallation) nor operating system match nor any add-ins. This also dispenses with any specific qualities of the user's machine (such as computer or smartphone). Being on the Internet facilitates collaboration among researchers (with passwords, if needed). We have advocated web-based work in many instances ([2] or [3] or [4] or [5], in almost 20 years).

The present Statistics problem is also illustrative and a template, addresses the behavior of the (random) area and perimeter of polygons that are randomly generated and inscribed in a circle. A description is briefly given in the next Section about these two random variables.

The described operation of a web-based computation is the informatic problem addressed, for which no related resolution was found on the Web (apart from general information). The procedure will be clarified on its architecture and web use, with surprising results, deserving additional development.

In the title of one of our previous articles [7], we mentioned "simple science": indeed, nowadays, in our technological age (frequently, people, fashionably, even mistake "technological" for "informatic"), uncountable —indeed, most—articles in the technical and scientific publications describe their (possibly) excellent results, but (we have found no exception) do not give us any real confirmation of their programming. Some publications nowadays offer their programs, yet, in languages we do not know or in operating system we do not use. We always cite Wilkinson's 'FAIR' initiative [8]: "There is an urgent need to improve the infrastructure supporting the reuse of scholarly data." The tens of authors of this proclamation duly mention *data*, but the situation is worse regarding *programs* in the literature in science and technology.

In the Sections that follow: "Illustrative problem", the Statistics problem and its Mathematical basis are explainde; "Web applications" are briefly discussed, in view of their alleged paucity; and "Results, discussion" show the architecture of our computing, with practical results. Finally, some "Conclusions" are put forward.

---

[1] "web page" is adopted (instead of "webpage", "web-page", etc.) as in the online Cambridge Dictionary [1].

2   Illustrative Problem

This study addresses the behavior of the area and the perimeter of random polygons, of which, so as to make the present problem tractable (and without real loss of generality), only polygons inscribed in a circle are considered. Also, the order of the entities "area, perimeter" means that the area is the main objective of the study, but the trivial calculation of the perimeter is included. It is the Cartesian coordinates of the vertices of a polygon that will completely define it.

The calculation of the area of any simple polygon (simple, *i.e.*, not star-shaped[2]), be it convex or not, seems complicated to a non-specialist. Surprisingly simple formulas, however, directly solve the problem, since the time and work of Meister [9], in the XVIII century, later consolidated by C. F. Gauss and C. G. J. Jacobi, in the XIX century. Meanwhile, the choice of polygons inscribed in a circle led to the option to sort the subtending angles, thus, easily excluding star polygons, and (unnecessary) making the polygons convex.

Various polygons are exemplified: Fig. 1 illustrates "random" polygons (indeed, as might have been randomly generated) inscribed in a circle, which are the target of this article; and Fig. 2 shows a concave polygon and a star polygon. These latter are out of our scope: concave polygons obviously cannot be inscribed in a circle, although the formulas below fully apply; and star polygons do not correspond to an area as considered here.

For the interested reader, Fig. 3 shows several polygons that were made at a free access web page of ours, [11]. Let us remark that, although generating a random polygon inscribed (or not) in a circle is very easy, the task becomes a complexity hard problem if, namely, the polygon is required to be "simple" (*i.e.*, non self-intersecting). To avoid this intricacy (out of our scope here), we had recourse to sort the vertices along the circle (increasing angle θ, Eq. 2), *i.e.*, 'Angle sorting' Yes, in the web page. (The coordinates of the vertices can be listed by enabling the 'Show values' option.)

One of various simple formulas [10] to calculate the area of a polygon with n vertices, from the Cartesian coordinates of its vertices, is

$$A = \frac{1}{2} \left| \sum_{i=1}^{i=n} (x_i y_{i+1} - x_{i+1} y_i) \right| \tag{1}$$

where (x_i, y_i), $i = 1..n$, are the coordinates, with the complementary condition of $(x_{n+1}, y_{n+1}) \equiv (x_1, y_1)$. (The absolute value is necessary because the value depends on the selected "direction" of the vertices.) As the polygons are inscribed in a circle, their coordinates are, of course,

$$(x_i, y_i) = \rho(cos\ \theta, sin\ \theta) \tag{2}$$

---

[2] Star polygons are also called self-intersecting.

where the radius of the circle, ρ, will be made unitary, and the θ's are random angles in $(0, 2\pi)$. The calculation of the perimeter is trivial, knowing that the angles have been sorted. Of course, the area, A, and the perimeter, P, of the random polygons must be in the following intervals, for a radius $R \geq 0$, for the number of vertices going from 2 to infinity.

$$4R \leq P \leq 2\pi R \tag{3}$$
$$0 \leq A \leq \pi R^2$$

Fig. 1. "Random" polygons: a quadrilateral ($n = 4$), and a pentagon ($n = 5$).

Fig. 2. "Random" polygons: a concave pentagon, and a star quadrilateral.

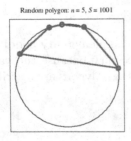
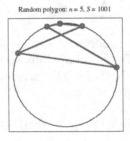

Fig. 3. Miscelanious "random" polygons with various n (RNG seed≡1001): (left to right) 5; ditto, sorted; 10.

In another Section, "Results, discussion", we delve into the computation, both regarding its Mathematics and statistical stages and its programming in the Internet "computing" medium. This is made accessible, freely and computable at the dedicated web page [12]. A number of web pages dealing with different subjects are ready and available.

3   Web and Applications

One of our principal objectives is to present an instance of our constructed web-based applications, by solving an illustration done through a Statistics problem, including the applicability to the work by students. We will comment on: Web, a medium for effective computing; students and computing; and the combination of different computer languages, each with its own characteristics.

3.1   Web and Computing

We frequently refer the singular example of V. M. Ponce (a professor of Hydraulics), whose website invites computations by the user, though other approximating websites exist. His site contains (since 2004) a virtual laboratory, "Vlab" ([13], through "OnlineCalc"), his perspective being similar to ours (since 1998). An example of his is described in his Problem No. 281, "Muskingum-Cunge convergence ratios", chosen here because of his (last in Google Scholar) publication in 2020 [14]. We show its input output in Fig. 4 (supposed input, as Hydraulics is out of our field). There, the user just inserts input data and receives output results. No default data, however, are supplied. Possibly, this is intended to make the user, typically the students, have to study and know what to fill in the fields, as commented next.

Academically, web pages such as ours and Ponce's do not weaken the need to study. We have observed that many students no longer do any programming—not their fault, if the professors follow the same careless path. Then, Excel becomes the modest fallback. If a web page solves a problem, students should do their own work (programming), and compare their results, such as we present ([12] once more). Many of our examples appeared in conferences (e.g., [6], on a chemical reaction, or [15]).

Worse than students' lack of interest is the fact that authors in numerical topics give no web pages with their computation for reproduction and confirmation. We have discovered none, and gave some negative samples in [5].

3.2   The Combination of Different Computing Languages

The combination, if advantageous, of different computing languages is much facilitated by current standards. Here, we use Python, Fortran, and C, with 'gnuplot' for graphing, under PHP, all free versions. Python is friendly to write a script and popular among scientists, and can act as "glue" ([16] or, e.g., [17]). It is (mailny) *interpreted*, so naturally slow. Fortran and C are *compiled*, therefore,

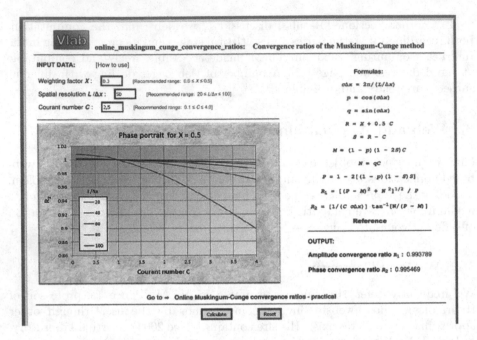

Fig. 4. Ponce's Problem (in Hydraulics) on "Muskingum-Cunge convergence ratios".

fast for number crunching. The present problem favors long run times, which is a characteristic in Monte Carlo applications, and this led to use fast languages. The present study turns into a template for hard problems, which is one of its objectives, as we keep suggesting the Web for scientific computing.

The language versions are: Python 3.9; Fortran 90, by GNU (Debian); C, Debian 10.2.1-6; 'gnuplot', 5.4; and PHP, 5.6. These versions are those installed in the public (Linux) system in the University.

The overall structure is a PHP web page that starts a Python script, which makes the computing but also invokes Fortran and C executables. About connecting Python to Fortran, it was done through the 'f2py' utility (as 'f2py3'), v. 2. This utility is able to convert a Fortran procedure (one or more subroutines) easily into a Python module, which is a "shared object", '.so' (or "dynamic linked library" in Windows terms). For instance, if 'f2py' is applied to a file in Fortran, 'NameF.f90', to produce 'NameF.so', the corresponding 'f2py3' produces (particularly in our Linux system) a 'NameF.cpython-37m-x86_64-linux-gnu.so', through

```
$ f2py3 —c —m NameF NameF.f90
```

In the python script itself, this module is used simply as 'import NameF'.

Regarding the connection of Python to C, the procedure was to use the Python 'ctypes' module. Previously. if the C file is 'NameC.c', the first steps are to create the 'NameC.so' through

```
$ gcc —c NameC.c
$ gcc —shared —o NameC.so NameC.o
```

Once the '.so' files have been created, they can be used by Python. In order to save space, the Python source is given as a downloadable annex in the problem web page (which will be seen at [11]).

4 Results, Discussion

As this study is directed to the Web, it does not show the computing itself. The task can be freely executed at

Ref. [12]

The user submits the problem data (input)—essentially, only the number of vertices—and obtains the desired results (output), which is the behavior of the area and perimeter. This simple scheme (aided by intrinsic default data) is, we think, adequate to a simple style in Engineering. The dynamic results come in HTML built with the Web-native scripting language PHP [18]. The architecture, in simple terms, has the following two elements: a front-end in the Web, to supply the interface to the user, and a back-end implementing the computing parts.

The Web interface is accessed by the user as the mentioned front-end, accepting the arguments of the task (data), processing them, scheduling the task, and which sends these data to the back-end, which, in turn, finishes the task, and replies to the front-end, thus providing the output built in a new (dynamic) web page. In technical problems, the output is here and often alphanumeric and or with graphics.

The above structure is used by us in our problems in Engineering, Mathematics and Statistics, because our goal is to solve them and make them freely available, in order to stress the proposal of this perspective of computing.

The numerical parts of the present problem are resolved in Python, Fortran, and C. The graphical functionalities exist in Python through the use of packages ('modules') such as 'matplotlib' or 'seaborn', whereas Fortran and C have no equivalent. From our previous frequent usage, and to establish a personal standard, we chose to use 'gnuplot' [19], a free graphical package, which can be neutrally called by many languages. Otherwise, for us, 'gnuplot' is more familiar than Python's own plotting libraries. On the Web, the computed results are cast in the HTML format (essentially using the HTML tag 'pre', meaning pre-formatted text), and displayed on the user's terminal.

The constructed computing solver is executable on our public system, maintained by CIIST, the Computing Centre of the University [20]. It is a Linux operating system and is characterized by 16 GB memory; the 'uname' command reports amd64, Debian 5.10.70 (2021), x86_64 GNU/Linux; and the 'lscpu' command, 8 CPU's, Xeon, 2 GHz.

The architecture can be described in the following steps, where the names (in bold, italics) are those of the particular case:

a. The user reaches the web page where the problem is shown (PHP), named ***P-RandPolygons.php***, and introduces the adequate data (input) in an HTML 'form' fields in the page. The data are sent to the next file through an HTTP request ("P-" here recalls that the engine is a Python script).
b. When clicking the 'Execute' button, the problem data are transmitted to an intermediate file, also in PHP, making use of the 'form' statement 'action=***RandPolygons.php***'.
c. The above mentioned PHP file takes the input received, preparesit, and sends it to a Python script (next). The transmission is as command line arguments for execution. The PHP file takes the results from the execution, via *stdout* and makes a (temporary, dynamic) PHP file (web page) that contains the results.
d. The standalone Python script, ***randPolygons.py***, is now run through a PHP 'exec' statement. The script in consideration depends on the user's choice: it runs the whole calculations (in Python); or passes parts of the computing to a Fortran subprogram, via module ***modRandPolygons***; or passes, ditto, to a C subprogram, using a 'ctypes' connection to the shared object ***simulate_polygons.so***.
e. The output from Python (to *stdout*) is sent to the Results page, a temporary and dynamic page. It contains an HTML 'pre' tag with the text and or graphics ('png'), which were produced through a call to the Python 'base64' procedure. Thus, this assures that no files stay on the server, which would need deletion. Importantly in public access, it also prevents clashes that would occur between various users using the web page.
f. Auxiliary files—The PHP characteristic environment files inserted by 'include', *i.e.*, and cascading style sheet and the proper images that identify the website itself.

The essential files needed to implement the actions mentioned are also listed in the web page for conciseness.

In the description above, the file names (emphasized for clarity) are the ones that depend on the specific problem. Vaious other files are common to other (possibly many) problems. In Fig. 5 is schematically shown the structure, detailed in item 'd.' in the list of steps displayed above.

The circumstance of running "over the Web" does not increase the difficulty to the usual resolution of a scientific or technical problem, where the ineluctable, hard task is to know the scientific subject.

The basis language that we have used is Python. After selection by the user, it runs alone or calls Fortran or C, and always 'gnuplot' for the plotting. Writing a program in languages such as Fortran or C is time consuming by comparison to popular user-friendly Matlab or Python (or Julia, for that matter). Notwithstanding, as regards execution, the difference in run time can turn out to be

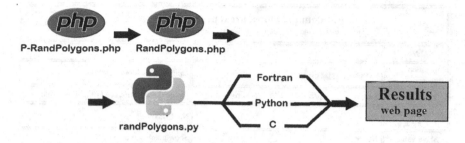

Fig. 5. Schematical network showing the communication among the web page components.

substantial, which favors the former languages (Julia, *e.g.*, [21], claims to solve this challenge). On an open environment as the Internet (and always), licensing must also be considered in using the languages (Fig. 6).

Our problem to observe the behavior of two random variables, area and perimeter, is now briefly described. The data for the example are the following (with the default values shown):

– Number of vertices (4);
– Simulation language (Fortran);
– Trials (10^6) and seed (75357) (for the Monte Carlo simulation);
– Histogram classes (200) (for the plots of the areas and perimeters);
– Show values (No) (of the plot coordinates).

Default simple data for the problem are, as we always do, given in the web page, permitting immediate results just with a click on the 'Execute' button.

The results of the execution are reproduced for increasing number of polygon vertices in the following figures. What was called "behavior" are, obviously, histograms of the simulated values of areas, A, and perimeters, P, or, finally in statistical terminology, "probability density functions" (pdf). Figure 7, for 3 vertices, shows the pdf's of the area, pdf$_A$, monotonically decreasing, and of the perimeter, pdf$_P$, with a surprising and abrupt peak at 4.0. In Fig. 8: for 4 vertices, the curve pdf$_A$ becomes a "cupola", and pdf$_P$ shows an abrupt peak at 5.2; and for 5 vertices, the curves become both smooth. Finally, in Fig. 9, the curves pdf$_A$ and pdf$_P$ acquire common shapes. These last shapes become intuitive because, as seen with many (12) vertices, a high probability is, as expected, for polygons close to the circle in which they are inscribed.

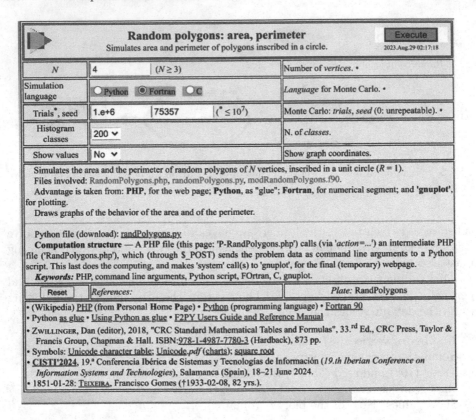

Fig. 6. Web page (input) for the random polygons' problem, with the default values.

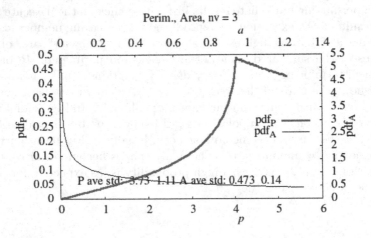

Fig. 7. Behavior of the area, A, and perimeter, P, of polygons of $n = 3$ ('nv' in images) vertices.

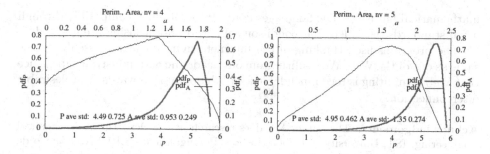

Fig. 8. Behavior of the area, A, and perimeter, P, of polygons of $n = 4$ (left) and $n = 5$ (right) vertices

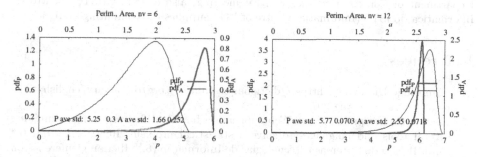

Fig. 9. Behavior of the area, A, and perimeter, P, of polygons of $n = 6$ (left) and $n = 12$ (right) vertices

5   Conclusions

In our opinion, the Internet, as a computing medium, has been a relatively forgotten field in science and technology. We use to cite one (rare) example by someone else, whereas, disquietingly, myriads of numerical publications permit no web-based confirmation. Our practice just needs a browser, avoiding any software installation by the user, who is able to easily supply the problem data. In the illustrative example, a statistical problem is solved, dealing with random polygons inscribed in a circle, with results that are surprising but intuitive in the asymptotic case for high number of vertices. Three languages, Python, Fortran, and C, are used to show the template nature for use in other problems, under PHP, with 'gnuplot' as graphing utility. The resulting web page is open to execution on the authors' website. In this study, we have meant to: recommend using this computing based on the client-server model, which greatly facilitates access without any specific need from the user; focus on students to bring them to computing; and to unveil and recall the benefits of bringing different languages together, with Python as a proficient candidate as a setting.

The problem studied reveals the Web as a suitable medium for scientific and technical computations. The architecture adopted serves an extensive spectrum of problems that, obviously, depend on the ability to tackle their unavoidable mathematical hardship. The languages used can be other, if compatible through usual standards, provided that conditions of speed and licensing are granted.

The programs used thinking of the Internet environment are essentially the same as out of the Web. We reaffirm from our academic and industrial experience that Web computing is recommendable for both domains, towards their desirable interconnection.

Acknowledgements. The author MC does research at the Department of Chemical Engineering, IST, University of Lisbon, Lisbon, Portugal, and CERENA, "Centro de Recursos Naturais e Ambiente" (*Centre for Natural Resources and the Environment*), under Project UID/04028/2020, funded by FCT, "Fundação para a Ciência e a Tecnologia" (Portuguese *National Science Foundation*); and PP is MSc student at the Department of Computer Science and Engineering, also of IST. CIIST, "Centro de Informática do IST" (Informatics Centre of IST) supplies the computing system.

References

1. Cambridge Dictionary. https://dictionary.cambridge.org/dictionary/english/webpag. Accessed 23 Aug 2023
2. Franco, B., Casquilho, M.: A Web application for scientific computing: combining several tools and languages to solve a statistical problem. In: CISTI 2011, 6.ª Conf. Ibérica de Sistemas e Tecnologias de Informação (6.th Iberian Conference on Information Systems and Technologies), Chaves (Vila Real), Portugal (2011)
3. Casquilho, M.: Computação científica, Internet, Indústria (Scientific computing, Internet, Industry), 1.st Portuguese Meeting on Mathematics for Industry, FCUP, University of Porto, Porto (Portugal) (2013)
4. Barros, M., Casquilho, M.: Linear Programming with CPLEX: an illustrative application over the Internet. In: CISTI 2019, Coimbra (Portugal) (2019)
5. Casquilho, M., Paredes, A., Rosa, F.C., Miranda, J.L.: A web-based cooling tower application. In: CISTI 2021, 16.ª Conf. Ibérica de Sistemas e Tecnologias de Informação (16.th Iberian Conf. on Information Systems and Technologies), Chaves (Vila Real, Portugal) (2021)
6. Casquilho, M., Paredes, A., Miranda, J.L.: Parameter estimation over the Internet for ODEs in a chemical reaction model. In: CISTI 2020, 15.ª Conferencia Ibérica de Sistemas y Tecnologías de Información (15.th Iberian Conference on Information Systems and Technologies), Sevilla (Spain) (2021)
7. Casquilho, M., Pires, A., Rosa, F.C., Miranda, J.L., Bordado, J.: A web-based superabsorbent polymer solver (in simple science). In: CISTI 2022, 17.ª Conf. Ibérica de Sistemas y Tecnologías de Información (17.th Iberian Conference on Information Systems and Technologies), Madrid (Spain) (2022)
8. Wilkinson, M.D.: Comment: the FAIR guiding principles for scientific data management and stewardship. Sci. Data **3**, 160018 (2016)
9. Meister, A.L.F.: Generalia de genesi figurarum planarum et inde pendentibus earum affectionibus. Deutsche Digitale Bibliothek (1769). https://www.deutschedigitale-bibliothek.de/person/gnd/100793118. Accessed 23 Aug 2023
10. Marsden, J.E., Tromba, A.J.: Vector Calculus. W. H. Freeman, New York (2003)
11. Casquilho, M.: Random inscribed polygon (2023). http://web.tecnico.ulisboa.pt/~mcasquilho/compute/explore/inscrPolygon/P-inscrPolygon.php

12. Casquilho, M.: Random polygons: area, perimeter (2023). http://web.tecnico. ulisboa.pt/~mcasquilho/compute/meeting/Polygons/P-RandPolygons.php
13. Ponce, V.M.: San Diego State University. http://ponce.sdsu.edu/. Accessed 23 Aug 2023
14. Ponce, V.M.: Muskingum-Cunge convergence ratios. https://ponce.sdsu.edu/ onlinemuskingumcungeconvergenceratios.php. Accessed 23 Aug 2023
15. Carolino, E., Casquilho, M., Rosário Ramos, M., Barão, I.: Applied scientific computing over the Web: robust methods in Acceptance Sampling for Weibull variables. ISBIS 2016 Meeting on Statistics in Business and Industry, 08–10 June, Barcelona (Spain) (2016)
16. The SciPy community. Using Python as glue. https://numpy.org/doc/stable/user/ c-info.python-as-glue.html. Accessed 23 Aug 2023
17. Oliphant, T.E.: Python for scientific computing. Comput. Sci. Eng. **9**(3), 10–20 (2007)
18. PHP, The PHP Group. https://www.php.net/. Accessed 23 Aug 2023
19. Gnuplot. http://www.gnuplot.info/. Accessed 23 Aug 2023
20. CIIST. Centro de Informática do IST. (Informatics Centre of Instituto Superior Técnico). https://ciist.ist.utl.pt/eng.php. Accessed 23 Aug 2023
21. Julia. http://julialang.org/. Accessed 23 Aug 2023

Hybrid Techniques

Multi Disease Prediction Using Ensembling of Distinct Machine Learning and Deep Learning Classifiers

M. Chaitanya Datta, B. Venkaiah Chowdary, and Rajiv Senapati[✉]

SRM University, AP, Amaravati, India
{chaitanyadatta_m,venkaiahchowdary_b,rajiv.s}@srmap.edu.in

Abstract. Diabetes, often regarded as a chronic illness, is a condition that occurs due to high blood sugar for a prolonged period of time. The risk of obtaining diabetes can be reduced by precise early prediction and analysing factors such as hereditary involvement and several other factors. Although advanced techniques came into existence, we can observe that the risk of developing diabetes is substantially higher among adults due to modern life. Timely treatment and diagnosis are required to prevent the outbreak and the advancement of diabetes. The lack of robustness in the precise early prediction of diabetes is a rigid task due to the size of the dataset and deficient labelled data. In this literature, we propose an architectural framework for the early prediction of diabetes disease where data pre-processing, outlier detection and avoidance, K-fold cross validation, and distinct predictive machine learning (ML) and deep learning (DL) classifiers (Decision Tree, Logistic Regression, and Neural Network) are appointed. In this literature, the ensembling of various machine learning and deep learning classifiers are used as a method of enhancing diabetes prediction, utilising K-fold cross validation as a validation strategy. The base classifiers are hypertuned using the grid search approach by considering numeric hyperparameters. The experiments conducted in this literature were conducted under similar conditions using the benchmark PIMA Indian Diabetes (PID) dataset. As a substitute for the conventional approach of testing the proposed approach, we have chosen the chronic kidney disease (CKD) dataset from the University of California (UCI) machine learning repository as a comparative study.

Keywords: Machine Learning · DBSCAN · Ensemble Model · k-fold cross validation · Multi disease prediction

1 Introduction

Diabetes is the most known disease in the world and has been facing crucial challenges to treat in all countries [12]. The pancreas is responsible for producing the insulin hormone, which helps the glucose to move from the intestine

Data Science Lab. SRM University, AP.

into the blood stream. The insulin deficiency can result in abnormal functioning of the pancreas, which creates diabetes, which can cause several conditions in the human body such as coma, weight loss, beta cells, and a lot more [29]. The condition can result in impactful changes in the behaviour of an individual and might cause damage to their capital and lifestyle if not treated at an early stage. As per the World Health Organisation (WHO), there exists 422 million people who are identified as diabetic worldwide, with the majority living found in middle-income and low-income countries. There are 1.5 million deaths directly referenced to diabetes every year [1]. The main motive behind proposing this approach is the need for an affordable system capable of assessing patient parameters and determining the existence of diabetes.

In this literature, we followed multiple classifier systems, treated as ensemble learning systems, which came into existence and have been applied to a huge set of problems [7]. Ensembling is a mechanism for integrating several learners together, and this can aid in improving the prediction capability and the performance of the proposed approach [2]. The technique of combining distinct predictive classifiers is called ensemble learning. They have been combined to effectively address a wide range of machine learning issues in the real world, including incremental learning and feature selection. This type of learning helps improve the confidence of a classifier by assigning weights to several individual classifiers and integrating them to achieve an outcome. In our paper, we used a voting technique to predict diabetes using ensemble learning. The predictive base classifiers used for ensemble learning are Decision Trees (DT), Logistic Regression (LR), and Neural Networks (NN). The base classifiers are chosen distinctly based on their diverse nature to achieve peak performance [31].

Rest of this paper is organized as follows. The literature available is presented in Sect. 2, Materials and methods presented in Sect. 3, The ensembling of distinct ML and DL classifiers are presented in Sect. 4, Results and discussion is presented in Sect. 5, and Finally, Sect. 6 concludes this paper.

2 Literature Review

Recent research has revealed that a sizable amount of relevant work has been documented in diabetes prediction by executing a variety of approaches to achieve peak performance; a few of them were presented in this literature. In [33], they predicted the diabetes using machine learning algorithms (DT, naïve bayes (NB) and Random Forest (RF)) using Hadoop cluster and map reduce techniques, and they achieved an accuracy score of 94% using RF. The experiment is conducted on a dataset gathered from the National Institute of Diabetes, consisting of data gathered from 75,664 patients with 13 attributes. In [28], they used distinct machine learning approaches such as NB, DT, support vector machine (SVM), and artificial neural networks (ANN) for the prediction of diabetes using the PID dataset. The experiment was conducted on a 75:25 ratio for training and testing the proposed approach and achieved an accuracy score of 82% for support vector machine and ANN. In [5], they followed a weighted ensembling approach using machine learning classifiers (k-nearest neighbour (KNN), DT, RF, adaboost, and multilayer perceptron (MLP)) on the PID dataset and achieved better results by experimenting with several ensemble models with different machine learning classifiers as base classifiers, hypertuning the hyperparameters of the base classifiers to achieve best performance using grid search, and cross-validating them. They have cross-validated their experiment using k-fold cross-validation (KCV) with k = 5. In [13], they pre-processed the dataset and clustered the dataset into two clusters using the k-means clustering technique, and as a predictive approach, they used a pipeline mechanism of having several machine learning classifiers. Their experiments are evident in getting better accuracy and results using the clustered dataset from the PID dataset, and the pipelining mechanism using Adaboost got an accuracy of 98.8%. In [6], the authors proposed an unweighted majority voting ensemble approach using machine learning classifiers (LR, KNN, gradient boosting, and RF) to predict diabetes. The dataset used in conducting their experiment is from National Health and Nutrition Examination Survey (NHANES) 2013–14, which has 10,172 data records with 54 features, and they achieved an area under the curve (AUC) score of 0.75. Some of the other approaches available in the literature [3,8–10,14,16,18,20,22–26,32] can also be used in the medical domain for various applications. The availability of quality medical data and efficient ML model for processing those data is always a challenging issue. In this paper, an ensembling model based frame work is proposed and the performance of this proposed model is evaluated using PID and CKD dataset.

3   Materials and Methods

The PID dataset and CKD dataset [CKD] are used in conducting this experiment, which consists of 400 and 768 data samples, respectively. The class feature and outcome feature indicate the existence or non-existence of the disease. We created several plots for a better understanding of the discrete and continuous data-valued attributes. In this paper, a diverse combination of machine learning and deep learning classifiers are considered as base classifiers in constructing the unweighted ensemble model—DT, LR, NN. The model is trained and applied with the voting rule to achieve better prediction accuracy, and the architectural framework of the proposed ensemble approach is represented in Fig. 1.

3.1   Dataset

In conducting this experiment, the PIMA Indian Diabetes (PID) Dataset [27] and chronic kidney disease dataset [21] are used. The PID dataset is gathered from 768 female diabetic patients sourced from the Pima Indian population residing in Arizona. The PID dataset is comprised of 768 patient records with 8 attributes. The outcome or class variable determines a particular patient record with a set of parameters that exhibit the existence or non-existence of the disease. The CKD dataset is collected from the University of California (UCI) machine

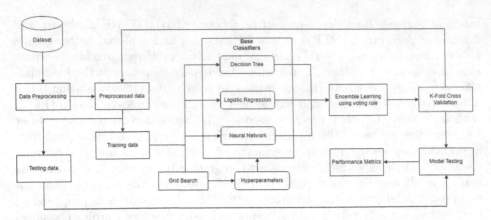

Fig. 1. A systematic representation of the proposed ensemble model for multi disease prediction.

learning repository. The CKD dataset is used for comparison of the proposed approach and to build evidence regarding how the size of the dataset effects the performance of the model. Ensembling strengthens the base classifiers by achieving great performance by ignoring the size of the dataset.

3.2   Data Preprocessing

The pre-processing of PID diabetes data is performed by collecting several patient records. The dataset involves nine attributes with a summary of 768 female patient data records, and we found a few instances among those records that exhibit signs of missing or null values. The missing values are filled using the mean and median of respective attributes. The data pre-processing is done by transforming all the medical records stored as diagnostic values. The results obtained post-processing show that out of 768 medical patient records, 500 data records were shown as 0, indicating non-diabetic patients, whereas the other 268 data records showed a value of 1, representing diabetic patients, which are exhibited in Fig. 2.

Outlier Detection and Analysis. Outliers are often considered noise among the hidden data and need to be removed or identified using an analysis of the data distributions among attributes. This may cause the model to see a sudden change in the data and may lead to inaccurate models, which can lead to inappropriate values being distributed between them. In our proposed approach, outliers are identified among the data points using pair plots as it is presented in Fig. 3. The outliers identified are removed from the dataset to achieve better performance. The correlation among the attributes is observed after the treatment of outliers, and the outcome is displayed in Fig. 4.

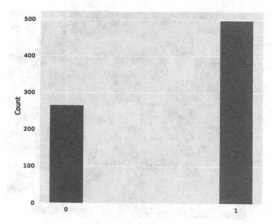

Fig. 2. The plot indicating the data points is distributed between non-diabetic and diabetic patients.

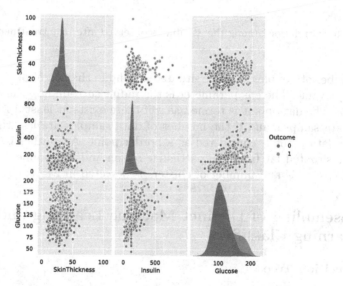

Fig. 3. The pair plots among the attributes (skinthickness, insulin and glucose) corresponding to the outcome variable where the outliers are found.

Density-Based Spatial Clustering of Applications with Noise (DBS-CAN). In the outlier analysis, DBSCAN [4] is used for the detection of multivariate outliers present among the data points. In this phase, the training data is considered, and the data points are grouped into multiple clusters for detecting outliers. The core objective of this approach is to find the dense areas belonging to a certain region by knowing a few parameters such as epsilon(eps) and minimum points($MinPts$), in which the data samples are divided into the core point (x), border point and an outlier. The minimum points parameter $MinPts$ is the

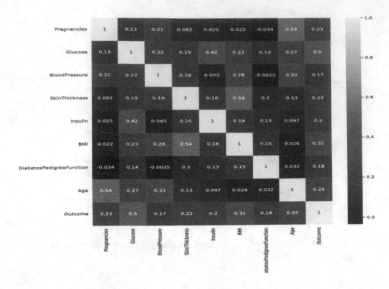

Fig. 4. The correlation among the attributes observed after the treatment of outliers.

least number of samples that are situated around in the neighbourhood in a particular *eps* value. The *eps* parameter is the radius surrounding the data samples around *x*. The data point is termed as *x* if there exists at least a *MinPts* number of data samples and if the number of data samples are less than *MinPts*, then the data point is considered as a border point. An outlier is a data point those are situated far from the particular region and the boundary. The graph between *eps* value and number of clusters is depicted in Fig. 5.

4   Ensembling of Distinct Machine Learning and Deep Learning Classifiers

4.1   Decision Tree

A Decision Tree (DT) [19] is a tree approach towards solving machine learning classification tasks. The trees are constructed using a top-down recursive approach using divide and conquer (DAC). The presence or absence of diabetes is predicted by developing a tree structure in which intermediate nodes indicate a feature, each branch defines an outcome of the feature, and each terminating node impacts an existence or non-existence of diabetes. The outcome label helps in identifying the diabetes. Consider a dataset with N classes. The formula can be used to calculate entropy.

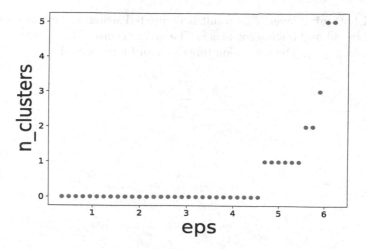

Fig. 5. The graph between the eps and number of clusters using DBSCAN for outlier detection.

$$Entropy(E) = -\sum_{i=1}^{m} p_i log_2 p_i$$

where p_i is the probability of randomly selecting an example in class 'i'.

4.2   Logistic Regression

Logistic Regression (LR) [30] is a supervised approach among the machine learning classification approaches which is found evident in solving several chronic diseases [15]. This approach estimates the relation between the independent variables $X = x_1, x_2, \ldots, x_n$ and a binary dependent variable (i.e., outcome variable) by estimating the probabilities using a sigmoid/activation function. This approach is applied only to a binary dependent variable (Y) with classes $0, 1$ and an independent variable that can be ordinal, ratio-level and binomial. The logistic/sigmoid function is mathematically given as:

$$\sigma(y) = \frac{1}{1 - e^{-y}}$$

where $y = w.X + b$ and $w = [w_1, w_2, \ldots, w_n]$ are the weights or coefficients, b is the bias or the intercept.

4.3   Neural Network

Neural Networks (NN) [11] involve a series of interconnected nodes in each layer, which consists of components such as inputs p_i and output q_i with a dropout layer and flattening layer. The result is computed using an activation function such as sigmoid and a constant bias b. The architecture of the neural network is presented in Fig. 6. The activation function can be presented as:

$$f\left(b + \sum_{i=1}^{n} p_i q_i\right)$$

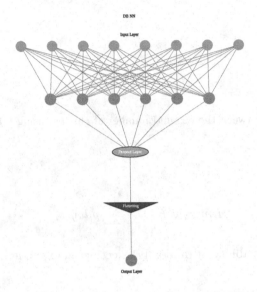

Fig. 6. The architecture of the neural network with several layers used in developing the proposed approach.

4.4   Combining the Base Classifiers Using Voting Rule

The outcome of an ensemble model [17] is dependent on the outcome of the base classifiers used in developing the ensemble model. This technique binds the probabilities of each prediction for each model and considers the prediction with the highest total probability. The integration of several distinct classifiers can aid in achieving better performance with the help of the soft voting rule. In this study, the soft voting rule applied, where m_i be the ith classifier of an ensemble model having N distinct classifiers which are set to predict the target class from a set of k possible classes $c_1, c_2, \ldots, c_n$ (in our study it's binary). For a provided input x, the classifier m_i will result in a k dimensional vector $[m_i^1(x), m_i^2(x), \ldots, m_i^k(x)]$, where $m_i^j(x)$ is an outcome of the ith classifier for the jth class label. All N distinct classifiers are considered with same priority unless weights are assigned for each classifier. The final classification is resulted by averaging the individual outcomes with different class label and then choosing the label k^* with respect to the highest majority vote (hard voting). The proposed ensemble model considers the class from a total of j classes that results in a maximum probability and combines them by adding all the probabilities of each class predicted by i^{th} classifier.

$$k^* = argmax \frac{1}{N} \sum_{i=1}^{N} m_i^j(x) - 1$$

In this soft voting rule, $m_i^j(x)$ is an estimate of the posterior probability $P(c_j|x)$ obtained by the ith base classifier.

5 Results and Discussion

As a part of the evaluation of the model, several measures must be employed to verify the model's performance and to find the necessary improvements that can occur to improve the overall performance. The evaluation is carried out with the confusion matrix, which has mainly four outcomes known as True Positive (TP), True Negative (TN), False Positive (FP), and False Negative (FN). The actual value refers to the existing value, and the predicted value refers to the value that is predicted by the proposed model. The measures employed for evaluation are as follows:

5.1 Accuracy

The most common measure to evaluate any model that finds how accurate the model is built. It can be termed as the ratio of the sum of correctly classified values to the sum of correctly classified values and misclassified values. Accuracy (ACC) is determined as $ACC = \frac{TP+TN}{TP+FN+FP+TN}$.

5.2 Precision

The measure of exactness or preciseness of the classification for evaluating the positively classified data values and is considered the ratio of the positively classified values to the sum of correctly classified data. $PRE = \frac{TP}{TP+FP}$ is used to calculate precision (PRE).

5.3 Recall

Recall is termed as the remembering/recollection/memorizing as a measure of quality and it is said to be the ratio of the positively classified data to the sum of correctly positive and negative classified data. Recall (or) True Positive Rate (or) Sensitivity (REC, TPR, SEN) is determined as $REC = \frac{TP}{TP+FN}$.

5.4 F1-Score

F1-Score is termed as the harmonic mean of the precision and recall to determine the average rate. The purpose of this metric is mainly used to evaluate imbalanced datasets, which do not have a balance among those classes between features. It works well on categorical attributes. As accuracy is not a good metric since it's more biased towards the positives rather than the negatives. F1-Score can compensate for them and hence gives an unbiased result while evaluating the model. F1-Score (F1) is determined as $F1 = 2 \times \frac{PRE*REC}{PRE+REC}$; where PRE, REC indicates precision and recall respectively. A comprehensive comparison between the base classifiers and the proposed ensemble approach using voting rule is depicted in Table 1, Fig. 7 and Fig. 8.

Table 1. A comparison between the base classifiers and corresponding metrics with the proposed ensemble approach.

Dataset	Classifiers	Accuracy	Precision	Recall	F1-Score
PID [27]	DT	75.65	63.16	74.07	68.18
	LR	80	77.78	60.49	68.06
	NN	71.3	70.27	32.1	44.07
	Ensemble model using hard voting rule (Proposed)	75.22	62	76.54	68.51
	Ensemble model using soft voting rule (Proposed)	74.78	84.85	34.57	49.12
CKD [21]	DT	98.75	100	98.08	99.03
	LR	92.5	100	88.46	93.88
	NN	77.5	100	65.38	79.07
	Ensemble model using hard voting rule (Proposed)	93.75	100	90.38	94.95
	Ensemble model using soft voting rule (Proposed)	98.75	100	98.08	99.03

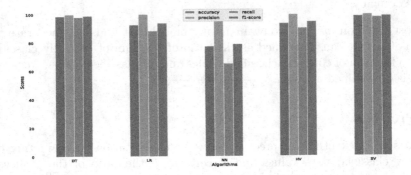

Fig. 7. A comparison among the proposed ensemble model and base classifiers with several performance metrics using CKD dataset.

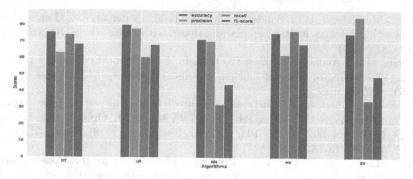

Fig. 8. A comparison among the proposed ensemble model and the base classifiers with several performance metrics using PID dataset.

6   Conclusion

In this work, ensembling aids in providing a better path towards the prediction of diabetes and chronic kidney disease using the PID and CKD datasets. The base classifiers are a diverse combination of machine learning approaches to solve the multi-disease prediction problems, which are integrated using a voting technique. The dataset was pre-processed using several techniques, which improved the overall model's performance and quality. The validation of the proposed approach was verified using 3-fold cross-validation (k = 3) among all the base classifiers and the proposed ensemble approach. The base classifiers are hypertuned using grid search by considering numerous hyperparameters of each base classifier to achieve peak performance. The comparative results make it evident that the proposed approach has outperformed various machine learning classifiers. Additionally, the proposed unweighted ensemble framework can be applied to related medical problems to prove its robustness and generality. This experiment has yielded convincing evidence that the performance of the model fluctuates by the scale of the dataset. By ensembling, we may reinforce the suggested technique even when the size of the dataset increases gradually.

References

1. https://www.who.int/news-room/fact-sheets/detail/cardiovascular-diseases-(cvds)
2. Akin Ozcift, A.G.: Classifier ensemble construction with rotation forest to improve medical diagnosis performance of machine learning algorithms. Comput. Methods Programs Biomed. **104**(3), 443–451 (2011)
3. Chakravadhanula, A.S., Kolisetty, J., Samudrala, K., Preetham, B., Senapati, R.: Novel decentralized security architecture for the centralized storage system in hadoop using blockchain technology. In: 2022 IEEE 7th International Conference for Convergence in Technology (I2CT), pp. 1–4. IEEE (2022)

4. Ester, M., Kriegel, H.P., Sander, J., Xu, X., et al.: A density-based algorithm for discovering clusters in large spatial databases with noise. In: KDD, vol. 96, pp. 226–231 (1996)

5. Hasan, M.K., Alam, M.A., Das, D., Hossain, E., Hasan, M.: Diabetes prediction using ensembling of different machine learning classifiers. IEEE Access **8**, 76516–76531 (2020)

6. Husain, A., Khan, M.H.: Early diabetes prediction using voting based ensemble learning. In: Singh, M., Gupta, P.K., Tyagi, V., Flusser, J., Ören, T. (eds.) ICACDS 2018. CCIS, vol. 905, pp. 95–103. Springer, Singapore (2018). https://doi.org/10.1007/978-981-13-1810-8_10

7. Lo, Y.T., Fujita, H., Pai, T.W.: Prediction of coronary artery disease based on ensemble learning approaches and co-expressed observations. J. Mech. Med. Biol. **16**(01), 1640010 (2016)

8. Datta, C., Senapati, R.: An adoptive heart disease prediction model using machine learning approach. In: 2022 OITS International Conference on Information Technology (OCIT), pp. 49–54. IEEE (2022)

9. Maddukuri, C.D., Senapati, R.: Hybrid clustering-based fast support vector machine model for heart disease prediction. In: Udgata, S.K., Sethi, S., Gao, X.Z. (eds.) International Conference on Machine Learning, IoT and Big Data, vol. 728, pp. 269–278. Springer, Singapore (2023). https://doi.org/10.1007/978-981-99-3932-9_24

10. Manda, S.C., Muttineni, S., Venkatachalam, G., Kongara, B.C., Senapati, R.: Image stitching using RANSAC and Bayesian refinement. In: 2023 3rd International Conference on Intelligent Technologies (CONIT), pp. 1–5 (2023). https://doi.org/10.1109/CONIT59222.2023.10205634

11. McCulloch, W.S., Pitts, W.: A logical calculus of the ideas immanent in nervous activity. Bull. Math. Biophys. **5**, 115–133 (1943)

12. Misra, A., et al.: Diabetes in developing countries. J. Diabetes **11**(7), 522–539 (2019)

13. Mujumdar, A., Vaidehi, V.: Diabetes prediction using machine learning algorithms. Procedia Comput. Sci. **165**, 292–299 (2019)

14. Muttineni, S., Yerramneni, S., Kongara, B.C., Venkatachalam, G., Senapati, R.: An interactive interface for patient diagnosis using machine learning model. In: 2022 2nd International Conference on Emerging Frontiers in Electrical and Electronic Technologies (ICEFEET), pp. 1–5. IEEE (2022)

15. Nusinovici, S., et al.: Logistic regression was as good as machine learning for predicting major chronic diseases. J. Clin. Epidemiol. **122**, 56–69 (2020)

16. Patro, P.P., Senapati, R.: Advanced binary matrix-based frequent pattern mining algorithm. In: Udgata, S.K., Sethi, S., Srirama, S.N. (eds.) Intelligent Systems. LNNS, vol. 185, pp. 305–316. Springer, Singapore (2021). https://doi.org/10.1007/978-981-33-6081-5_27

17. Polikar, R.: Ensemble learning. In: Zhang, C., Ma, Y. (eds.) Ensemble Machine Learning: Methods and Applications, pp. 1–34. Springer, New York (2012). https://doi.org/10.1007/978-1-4419-9326-7_1

18. Prasad, G.G., Chowdari, A.A., Jona, K.P., Senapati, R.: Detection of CKD from CT scan images using KNN algorithm and using edge detection. In: 2022 2nd International Conference on Emerging Frontiers in Electrical and Electronic Technologies (ICEFEET), pp. 1–4. IEEE (2022)

19. Quinlan, J.: Induction of decision trees. Mach. Learn. **1**, 81–106 (1986)

20. Raviteja, K., Kavya, K., Senapati, R., Reddy, K.: Machine-learning modelling of tensile force in anchored geomembrane liners. Geosynthetics Int., 1–17 (2023)

21. Rubini, L.: Early stage of chronic kidney disease UCI machine learning repository (2015)
22. Sahoo, A., Senapati, R.: A Boolean load-matrix based frequent pattern mining algorithm. In: 2020 International Conference on Artificial Intelligence and Signal Processing (AISP), pp. 1–5. IEEE (2020)
23. Sahoo, A., Senapati, R.: A novel approach for distributed frequent pattern mining algorithm using load-matrix. In: 2021 International Conference on Intelligent Technologies (CONIT), pp. 1–5. IEEE (2021)
24. Sahoo, A., Senapati, R.: A parallel approach to partition-based frequent pattern mining algorithm. In: Udgata, S.K., Sethi, S., Gao, X.Z. (eds.) Intelligent Systems, vol. 431, pp. 93–102. Springer, Singapore (2022). https://doi.org/10.1007/978-981-19-0901-6_9
25. Samudrala, K., Kolisetty, J., Chakravadhanula, A.S., Preetham, B., Senapati, R.: Novel distributed architecture for frequent pattern mining using spark framework. In: 2023 3rd International Conference on Intelligent Technologies (CONIT), pp. 1–5 (2023). https://doi.org/10.1109/CONIT59222.2023.10205903
26. Senapati, R.: A novel classification-based parallel frequent pattern discovery model for decision making and strategic planning in retailing. Int. J. Bus. Intell. Data Min. **23**(2), 184–200 (2023)
27. Smith, J.W., Everhart, J.E., Dickson, W., Knowler, W.C., Johannes, R.S.: Using the ADAP learning algorithm to forecast the onset of diabetes mellitus. In: Proceedings of the Annual Symposium on Computer Application in Medical Care, p. 261. American Medical Informatics Association (1988)
28. Sonar, P., JayaMalini, K.: Diabetes prediction using different machine learning approaches. In: 2019 3rd International Conference on Computing Methodologies and Communication (ICCMC), pp. 367–371. IEEE (2019)
29. Vaishali, R., Sasikala, R., Ramasubbareddy, S., Remya, S., Nalluri, S.: Genetic algorithm based feature selection and MOE fuzzy classification algorithm on pima Indians diabetes dataset. In: 2017 International Conference on Computing Networking and Informatics (ICCNI), pp. 1–5. IEEE (2017)
30. Wright, R.E.: Logistic regression (1995)
31. Yang, L.: Classifiers selection for ensemble learning based on accuracy and diversity. Procedia Eng. **15**, 4266–4270 (2011)
32. Yerramneni, S., Vara Nitya, K.S., Nalluri, S., Senapati, R.: A generalized grayscale image processing framework for retinal fundus images. In: 2023 3rd International Conference on Intelligent Technologies (CONIT), pp. 1–6 (2023). https://doi.org/10.1109/CONIT59222.2023.10205834
33. Yuvaraj, N., SriPreethaa, K.: Diabetes prediction in healthcare systems using machine learning algorithms on Hadoop cluster. Clust. Comput. **22**(Suppl 1), 1–9 (2019)

Resource Management Through Workload Prediction Using Deep Learning in Fog-Cloud Architecture

Pratibha Yadav$^{(\boxtimes)}$ and Deo Prakash Vidyarthi

School of Computer and System Sciences, Jawaharlal Nehru University, New Delhi, India
yadavpratibha26@gmail.com

Abstract. Workload prediction involves forecasting the future resource demands and patterns for a computing system, such as cloud or fog infrastructure. It applies historical information and analytical methods to forecast future workload patterns. By anticipating workload patterns, resource managers can allocate resources proactively, optimize system performance, and ensure efficient resource utilization. This research introduces a Fog-Cloud specific workload prediction model based on time series analysis, utilizing Intuitionistic Fuzzified C-mean clustering and Long Short-Term Memory (IFCM-LSTM), with emphasis on CPU and memory prediction. First, IFCM clustering incorporates uncertainty, allowing for a more flexible representation of resource patterns. This is particularly useful in scenarios where resource utilization behavior is uncertain or hesitant. Next, the LSTM model effectively captures the intricate temporal relationships in the historical workload data collected from fog-cloud nodes. The model's performance has been assessed on a real workload dataset using three evaluation metrics: Root Mean Square Error (RMSE), Mean Squared Error (MSE), and Mean Absolute Error (MAE). These metrics offer valuable insights into the accuracy of the model's predictions.

Keywords: Cloud and Fog Computing · Internet of Things (IoT) · Time Series Forecasting · IFCM (Intuitionistic fuzzified C-mean Clustering) · LSTM (Long Short-Term memory)

1 Introduction

As the number of Internet of Things (IoT) devices continues to grow with voluminous increase in the data generated at the network edge, there is a growing need for accurate prediction of workloads [1] in the fog-cloud layer. It has become critical for efficient resource management and utilization. Workload prediction in the fog layer involves forecasting the resource demands and patterns at the edge nodes, considering factors such as CPU and memory utilization, network traffic, and application-specific parameters [2]. Accurate workload prediction allows for proactive resource allocation, dynamic scaling, and system performance optimization to meet the varying demands of the applications.

© The Author(s), under exclusive license to Springer Nature Switzerland AG 2024
K. K. Patel et al. (Eds.): icSoftComp 2023, CCIS 2031, pp. 258–269, 2024.
https://doi.org/10.1007/978-3-031-53728-8_20

The rise of IoT and the associated volume of data with big data analytics leverage in decision-making for organizations [3]. However, one of the most challenging tasks, experienced by researchers and academics, is evaluating and predicting time-series data. It refers to data points representing the changing metrics over time, with each point denoted as $X(t)$, where t ranges from 0 to n. Time series data is considered univariate when it involves a single variable, while multivariate time series data involves multiple factors. The term "multifunctional time series data" describes the collection of such multi-factor time series information. The fundamental principle of time series forecasting is to extract meaningful insights from data to predict future outcomes. In context of cloud/fog computing, time series analysis is indispensable for workload prediction, resource allocation, and system optimization. ARIMA and SVM frameworks have evaluated and forecasted the sequential data in [4]. In domains, such as resource management and performance optimization in cloud/fog computing, various algorithms to forecast time series data have been effectively implemented and demonstrated.

Fog computing offers various services, through a pay-per-use model, to its end users. Service elasticity, which refers to the system's ability to manage fluctuating workloads during peak times, plays a crucial role. It dynamically allocates or de-allocates computing resources to closely match the service demand of the current time interval. By employing service elasticity, fog-cloud enhances Quality of Service (QoS) metrics, including service availability, throughput, and response time. It also contributes to reducing power consumption and avoids resource under/over-provisioning. Auto-scaling mechanisms, classified as reactive or proactive, are used to implement service elasticity. Reactive approaches utilize rule-based triggers based on system resource utilization, while proactive mechanisms estimate future resource requirements based on past utilization. Deep learning algorithms, a subset of machine learning techniques, have recently gained prominence in addressing such challenges.

This work proposes and implements an IFCM-LSTM based model for workload prediction, emphasizing CPU and resource utilization. The highlights of the work are as follows.

1. Utilizing the temporal memory and sequence learning capabilities of LSTM, it captures complex temporal patterns and dependencies present in workload data more precisely.
2. Through rigorous experimentation on real-world datasets, it is illustrated that the proposed IFCM-LSTM based model outperforms conventional methods in terms of prediction.
3. This study helps in efficient capacity planning and improved system performance in Fog-Cloud computing environment. The findings have implications for optimizing resource utilization, enhancing scalability, and guaranteeing efficient resource management in fog-cloud systems.

The remainder of this paper is organized as follows. In Sect. 2, we discuss the related studies and their limitations and present our contributions. In Sect. 3, we introduce our proposed prediction model. Section 4 presents the proposed algorithm for resource prediction. The performance evaluation results and experimental setup are presented in Sect. 5, followed by the conclusions in Sect. 6.

2 Related Work

In recent years, Fog service providers have encountered two important challenges: ensuring service quality (QoS) and competing with limited resources. Statistical techniques have addressed these challenges, particularly in predicting future resource usage. This topic has attracted significant research attention in computer systems and resource management. This section reviews related work, encompassing various methodologies and techniques employed for predicting future resource usage. It includes studies focusing on both traditional statistical methods and emerging machine-learning approaches.

Lalitha et al. [5] employed a statistical model called ARIMA-ANN to forecast future resources in cloud environments. Zhao et al. [6] resource-efficient strategy called the selective offloading based on ARIMA-BP (ABSO) has been introduced to meet delay requirements and minimize the energy consumption of mobile devices. ABSO uses the ARIMA-BP approach to estimate the computational capacity of edge clouds and then design selective offload algorithms to determine offload strategies. Kumar et al. [7] combined neural networks and independent differential evolutionary algorithms to predict workloads. Their model can learn the best possible mutation strategy and the best cross-cutting rate. The experiment was carried out with the benchmark data sets of the HTTP traces of NASA and Saskatchewan servers for different prediction intervals. In [8], authors proposed a traffic prediction method based on a statistical model using the Poisson distribution in a sliding window to calculate the weight of observations. The normalized mean square error between predicted and observed values was compared using a real-world cloud computing dataset obtained from monitoring Drop Box usage to evaluate the proposed method. A separate study by Le et al. [9] has proposed a new monthly cloud mitigation distribution model for different regions of Japan. The study used five-year ERA interim meteorological database, provided by the European Centre for Medium-Term Forecasting (ECMWF). The cloud reduction is calculated based on the cloud liquid water content (CLWC) extracted from the ECMWF ERA interim database. The probability density function (PDF) of cloud mitigation is determined by using a curve fitting method.

Numerous machine learning algorithms have been employed for resource prediction. Gao et al. [10] proposed a method of predicting a certain time before the predicted time, to allow enough time for task planning based on anticipated workloads. They introduced a clustered workload prediction approach that classified tasks into multiple clusters and trained separate prediction models for each category to improve prediction accuracy. Guo et al. [11] considers factors related to mobility and utilizing LSTMs to predict workload. They have solved the problem of uneven resource usage in the edge data center (EDC). Their method includes EDC's user mobility and geographical location information, facilitates resource allocation, and improves resource utilization. In a study by Kim et al. [12], linear regression was implemented to estimate the total processing time of individual tasks on candidate MEC nodes (mobile edge computing). Subtasks are then assigned to certain edge nodes based on the observed state of each MEC node. They also developed a core cloud monitoring module. Peng et al. [13] propose a network traffic forecast method for variable sampling rates, using LSTM and machine learning. The sampling rate determining the accuracy of traffic forecasts can change in real time with the dynamic change of network traffic.

3   The Proposed Prediction Model

This work proposes a prediction model based on IFCM-LSTM to predict the CPU and memory requirements. The model is a combination of intuitionistic fuzzy C-Mean (IFCM) clustering and long-term short-term memory (LSTM). It starts with the application of IFCM [14] clustering to group similar resource utilization patterns, incorporating uncertainty and flexibility in cluster assignment. The clustered representation is then fed into the LSTM network, which captures temporal dependencies and predicts future resource workloads. The combination of IFCM and LSTM leverages the strengths of both the techniques, resulting in improved prediction accuracy and interpretability.

LSTM [15] captures temporal dependencies for accurate predictions, while IFCM provides interpretability through membership and non-membership degrees, which is calculated using following steps:

Step 1. The membership function (m_{jk}) is calculated as Eq. (1).

$$m_{jk} = \frac{s_{jk}}{\sum_{j=1}^{n} s_{jk}} \tag{1}$$

where, m_{jk} are the membership values and s_{jk} is the random values generated from uniform distribution having parameters (0, 1) of kth observation of the time series data y_k ($j = 1, 2, \ldots, n$) and $k = (1, 2, \ldots, c)$.

Step 2. The hesitation levels h_{jk} is calculated using Eq. (2). Equation (3) is used to calculate intuitionistic fuzzified membership (m_{jk}^*). The value of intuitionistic membership (m_{jk}^*) is stored in the matrix m_{old}.

$$h_{jk} = 1 - m_{jk} - \left(1 - \left(m_{jk}^{\alpha}\right)\right)^{\frac{1}{\alpha}}, \alpha > 0 \tag{2}$$

$$m_{jk}^* = m_{jk} + h_{jk} \tag{3}$$

Step 3. Using m_{jk}^* obtained in *Step 3*, the center of clustering $\left(v_j^*\right)$ is computed as in Eq. (4).

$$v_j^* = \frac{\sum_{k=1}^{c} \left(m_{jk}^*\right)^f y_k}{\sum_{k=1}^{c} \left(m_{jk}^*\right)^f}; j = 1, 2, \ldots, n. \tag{4}$$

where, $y_k (k = 1, 2, \ldots, c)$ represents time-series data, m_{jk}^*, ($j = 1, 2, \ldots, n; k = 1, 2, \ldots, c$) is intuitionistic membership values derived in Step 2, and $f > 1$ is fuzziness factor.

Step 4. Equation (5) alters the membership functions (m_{jk}) obtained in *Step 1*.

$$m_{jk} = \frac{1}{\sum_{j=1}^{n} \left(\frac{d_{jk}}{d_{ik}}\right)^{\frac{2}{(f-1)}}}; j = 1, 2, \ldots, n;$$
$$k = 1, 2, \ldots, c. \tag{5}$$

$$d_{jk} = \sqrt{\left(y_k - v_j^*\right)^2} \tag{6}$$

In Eq. (5) and Eq. (6), m_{jk}, is the fuzzy membership value of the kth observation for the jth cluster produced in Step 1, the Euclidean distance (d_{jk}) is evaluation of kth observation in the jth cluster center, f is the fuzziness factor and $v_j^*(j = 1, 2, ..., n)$ is the jth cluster center obtained in *Step 3*.

Step 5. Equation (2) and Eq. (3) are used to modify the hesitation values h_{jk} and the intuitionistic fuzzy membership variable m_{jk}^*. In a matrix, named m_{new}, the updated instuitionistic membership variables (m_{jk}^*) are stored.

Step 6. v_{jk} (non-membership) values are calculated using Eq. (7) and m_{jk}^* are the updated values of intuitionistic membership calculated in *Step 5*. The new intuitionistic non-membership attributes (v_{jk}) are stored in matrix called V.

$$m_{jk}^* + h_{jk} + v_{jk} = 1 \tag{7}$$

Step 7. If the condition specified in Eq. (8) is satisfied, the algorithm is terminated. Otherwise, the algorithm proceeds by setting m_{old} equal to m_{new} and then returning to *Step 3*. ε is the small positive integer.

$$m_{new} = m_{old} < \varepsilon \tag{8}$$

The output produced by IFCM algorithm, such as membership value and non-membership values, are concatenated with the time-series data as the input to the LSTM.

The Recurrent Neural Network (RNN) [16, 17] comprised of an input, hidden, and output layer (Fig. 1). Due to the issues of gradient vanishing and gradient exploding, back propagation faces difficulties in effectively learning patterns with long-term dependencies [18]. These challenges play a crucial role in the training difficulties encountered in recurrent neural networks [19–21].

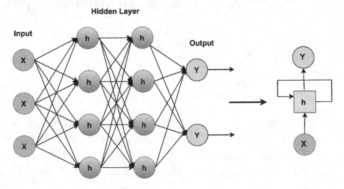

Fig. 1. Sequential Processing in RNN

Schmidhuber [22] introduced LSTM as an improved version of RNNs, aiming to overcome the issues of gradient vanishing and gradient exploding of conventional RNN. LSTM incorporate additional interactions within each module or cell, capable of learning long-term dependencies and remembering information for long period. Unlike standard RNNs, LSTMs consist of four interacting layers and employ a unique communication mechanism within their repeating modules. Figure 2 depicts the structural arrangement of the LSTM neural network which composed of blocks of memory called cells that transmit two states: cell state and hidden state. The cell state serves as the primary chain of the data flow, allowing data to forward essentially without any changes. However, the cell state is responsive to linear transformations, and sigmoid gates are used to add or remove data. These gates, like weighted layers, control the memory retention process in LSTMs and help address the problem of long-term dependencies.

The first step, in the development of LSTM network, involves the identification and discarding of unnecessary information from the cell at this stage. The sigmoid function is used to make the selection, considering the previous LSTM unit's output (h_{t-1}) of $t-1$ and the present input (X_t) of t. Furthermore, the sigmoid function determines the part of previous output that needs to be ignored.

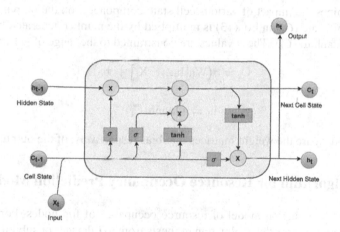

Fig. 2. Structure of the LSTM

f_t, called the forget gate, is represented by a vector of values ranging from 0 to 1, associated with each cell state element (C_{t-1}) shown in Eq. (9).

$$f_t = \sigma\left(W_f\left[h_{t-1}, X_t\right] + b_f\right) \tag{9}$$

σ represents the sigmoid function here while W_f and b_f stands for the bias and weight matrix of the forget gate (f_t).

The next step in the LSTM network is to determine and store information from a new input (X_t) and update the cell state in the following steps. The process consists of two parts: the sigmoid and tanh layers. Initially, the sigmoid layer decides whether to update or ignore new information and generates the binary values of 0 or 1. Next, the *tanh* function assigns weights to the selected values and determines its level of significance in the range of -1 to 1. The combination of these two values changes and updates the new cell state. The updated memory is then combined with the previous memory (C_{t-1}) to generate a new cell state (C_t) shown in Eq. (12).

$$i_t = \sigma\big(W_i[h_{t-1}, X_t + b_i]\big) \tag{10}$$

Equation (10) calculate input gate that serves the purpose of assessing the significance of the fresh data contained in the input. The new information is calculated using Eq. (11), that must be passed to the cell state is a hidden state function of the previous $t-1$ time value and the x time value. The activation function is Tanh.

$$N_t = tanh\big(W_n[h_{t-1}, X_t] + b_n\big) \tag{11}$$

$$C_t = C_{t-1}f_t + N_t i_t \tag{12}$$

C_{t-1} and C_t represent the cell states at time $t-1$ and t, respectively, W_i represents the cell state's weight matrix and b_i is bias. In final stage, the filtered output values (h_t) in Eq. (14) are obtained by applying a filter to the output cell state (O_t). Initially, a sigmoid layer determines the impact of various cell state components on the output. The result of the sigmoid gate (O_t) in Eq. (13) is multiplied by the number generated by applying the function tanh to (C_t). These values are constrained to the range of -1 to 1.

$$O_t = \sigma\big(W_0[h_{t-1}, X_t] + b_0\big) \tag{13}$$

$$h_t = O_t tanh(C_t) \tag{14}$$

Here, W_0 and b_0 are the weight matrices and bias respectively, of the output gate.

4   The Algorithm for Resource Occupancy Prediction Model

We designed a prediction model of resource occupancy of fog nodes. For Fog nodes they receive and process the application requests from IoT devices or submit application requests to cloud server to meet various quality of service requirements.

Input

$X_k\{k|k \quad 1,2,\ldots\ldots,N\}$(CPU occupancy datasets),n_c(no. of intuitionistic fuzzy sets), n_h(no. of hidden layers).

Result

Prediction result sets with status indicator predict result Y_k' .

1. Initialize data set collected by resource monitor.
2. Extract information used by the CPU separately.
3. Calculate the membership and non-membership values using algo 1.m_{jk}^* and $V_{jk}\{j = 1,2,\ldots,\aleph; k = 1,2,\ldots\ldots,N\}$.
4. Normalize the data and set the value range of data set X_k to (0,1)
5. **for** step k←1 to N **do**
6. Inputs and targets of the LSTM artificial neural network are prepared.
7. Calculate m_{jk}^* and V_{jk}.
8. Feed the dataset X_k, concatenate with m_{jk}^* and V_{jk} into the LSTM to process.
9. The LSTM is trained, and optimal weights are obtained.
10. Get the predicted result Y_k'.
11. **end for.**

5 Experiment Setup and Evaluation

In this section, the proposed model is evaluated on real-time data sets. The dataset is collected from 1,750 virtual machines (VMs) hosted in a distributed data center managed by Bitbrain [23], a service provider specializing in enterprise-focused managed hosting and business computation. The workload consists of two types of resource usage, CPU and Memory. 70% of the data is used in the training phase, and 30% for the testing phase. Experiments are conducted on an Intel Core i5-11300H CPU and 16 GB RAM. Python 3.10 and MATLAB 2020b are employed for data analysis and numerical computations, leveraging their respective strengths. The flow of the proposed model is shown in Fig. 3.

5.1 Evaluation Criteria

Absolute percentage errors (MAPEs), Average absolute errors (MAEs), and Average Square errors (MSEs) the three common metrics are used to evaluate the proposed model. These parameters give a complete understanding of the model's performance and help to choose the model best suited to the particular problem. In point value prediction, errors are determined by calculating the difference between actual and predicted workload, where y_i^p, y_i^a represent the predicted and actual value. Equations (15)–(17) define the MSE, MAE, and MAPE values as shown in Table 1.

$$MSE = \frac{1}{t+1}\sum_{1}^{t}\left[Error(i)^2\right] \tag{15}$$

$$MAE = \frac{1}{N}\sum_{i=1}^{N}(y_i^p - y_i^a) \tag{16}$$

$$MAPE = \frac{1}{N}\sum_{i=1}^{N}\left(\frac{y_i^p - y_i^a}{y_i^a}\right)*100 \tag{17}$$

Table 1. Prediction Error of the Model

S.No	Error	Values
1	RMSE	1.609
2	MAE	0.929
3	MAPE	0.166

The result provides an analysis of the usage of CPU and memory based on actual and forecasted data. Figures 4(a) and 4(b) show the current CPU and memory usage patterns on the Bitbrain database, providing insight into resource usage. Figures 5(a) and 5(b) show the CPU and memory used to predict and evaluate the accuracy of the prediction model. Fluctuations and potential increase in CPU and memory usage are identified, helping to understand the resource requirements. Overall, the section provides a comprehensive understanding of the observed and predicted resource usage patterns, which contributes to informed decision-making and potential improvements in performance.

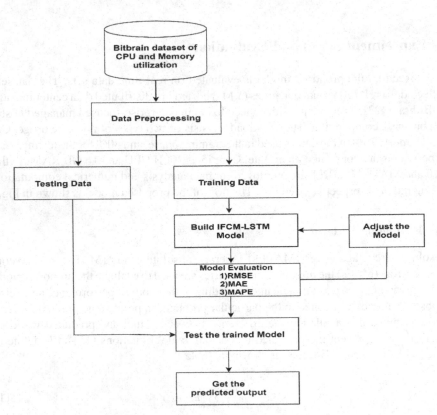

Fig. 3. Framework for the proposed model

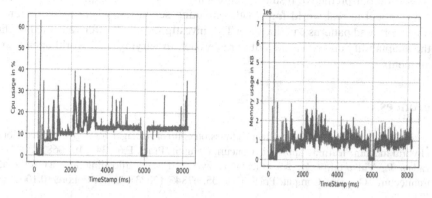

Fig. 4. (a). Sample Bitbrain data for CPU Usage. **(b).** Sample Bitbrain data for Memory Usage

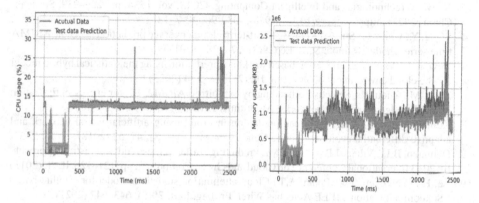

Fig. 5. (a). Model Prediction of CPU Usage. **(b).** Model Prediction of Memory Usage

6   Conclusion and Future Work

The work proposed the IFCM-LSTM model for predicting future CPU and memory utilization for fog-cloud nodes. The model utilized historical data on CPU and memory usage to forecast the future behavior of these resources. The prediction results show that the LSTM model accurately predicts future CPU use and memory. The accuracy of predictions is evaluated using indicators such as RMSE, MAE, and MAPE. These predictions can have practical applications in resource provisioning, workload balancing, and overall system performance optimization in fog-cloud computing architecture which results in making informed decisions and ensure the overall efficiency and reliability of the fog-cloud computing systems.

In the future, we plan to further enhance our research by incorporating dynamic offloading strategies, aiming to identify and route tasks to the most appropriate fog node based on our predictive results. Additionally, we intend to explore the practicality of our model in real-world fog-cloud computing scenarios, where network conditions and workload patterns can fluctuate. This investigation will offer valuable insights into the adaptability and robustness of our approach in varying, real-world operational environments.

References

1. Sham, E.E., Vidyarthi, D.P.: Intelligent admission control manager for fog-integrated cloud: a hybrid machine learning approach. Concurr. Comput. Pract. Exp. **34**(10), e6687 (2022)
2. Yadav, P., Vidyarthi, D.P.: An efficient fuzzy-based task offloading in edge-fog-cloud architecture. Concurr. Comput. Pract. Exp. **35**, e7843 (2023). https://doi.org/10.1002/cpe.7843
3. Yadav, P., Vidyarthi, D.P.: Analyzing the behavior of real-time tasks in fog-cloud architecture. In: Woungang, I., Dhurandher, S.K., Pattanaik, K.K., Verma, A., Verma, P. (eds.) Advanced Network Technologies and Intelligent Computing. CCIS, vol. 1534, pp. 229–239. Springer, Cham (2022). https://doi.org/10.1007/978-3-030-96040-7_18
4. Wang, Y., Wang, C., Shi, C., Xiao, B.: Short-term cloud coverage prediction using the ARIMA time series model. Remote Sens. Lett. **9**(3), 274–283 (2018)
5. Devi, K.L., Valli, S.: Time series-based workload prediction using the statistical hybrid model for the cloud environment. Computing **105**(2), 353–374 (2023)
6. Zhao, M., Zhou, K.: Selective offloading by exploiting ARIMA-BP for energy optimization in mobile edge computing networks. Algorithms **12**(2), 48 (2019)
7. Kumar, J., Singh, A.K.: Workload prediction in cloud using artificial neural network and adaptive differential evolution. Future Gener. Comput. Syst. **81**, 41–52 (2018)
8. Dalmazo, B.L., Vilela, J.P., Curado, M.: Predicting traffic in the cloud: a statistical approach. In: 2013 International Conference on Cloud and Green Computing, pp. 121–126. IEEE (2013)
9. Le, H.D., Nguyen, T.V., Pham, A.T.: Cloud attenuation statistical model for satellite-based FSO communications. IEEE Antennas Wirel. Propag. Lett. **20**(5), 643–647 (2021)
10. Gao, J., Wang, H., Shen, H.: Machine learning based workload prediction in cloud computing. In: 2020 29th International Conference on Computer Communications and Networks (ICCCN), pp. 1–9. IEEE (2020)
11. Guo, Q., et al.: Research on LSTM-based load prediction for edge data centers. In: 2018 IEEE 4th International Conference on Computer and Communications (ICCC), pp. 1825–1829. IEEE (2018)
12. Kim, K., Lynskey, J., Kang, S., Hong, C.S.: Prediction based sub-task offloading in mobile edge computing. In: 2019 International Conference on Information Networking (ICOIN), pp. 448–452. IEEE (2019)
13. Peng, R., Fu, X., Ding, T.: Machine learning with variable sampling rate for traffic prediction in 6G MEC IoT. Discrete Dyn. Nat. Soc. **2022** (2022)
14. Xu, Z., Wu, J.: Intuitionistic fuzzy C-means clustering algorithms. J. Syst. Eng. Electron. **21**(4), 580–590 (2010). https://doi.org/10.3969/j.issn.1004-4132.2010.04.009
15. Nitesh, K., Abhiram, Y., Teja, R.K., Kavitha, S.: Weather prediction using long short term memory (LSTM) model. In: 2023 5th International Conference on Smart Systems and Inventive Technology (ICSSIT), pp. 1–6, January 2023. https://doi.org/10.1109/ICSSIT55814.2023.10061039
16. Rumelhart, D.E., Hinton, G.E., Williams, R.J.: Learning representations by back-propagating errors. Nature **323**(6088), 533–536 (1986)

17. Werbos, P.J.: Generalization of backpropagation with application to a recurrent gas market model. Neural Netw. **1**(4), 339–356 (1988)
18. Bengio, Y., Simard, P., Frasconi, P.: Learning long-term dependencies with gradient descent is difficult. IEEE Trans. Neural Netw. **5**(2), 157–166 (1994)
19. Hochreiter, S.: The vanishing gradient problem during learning recurrent neural nets and problem solutions. Int. J. Uncertain. Fuzziness Knowl.-Based Syst. **6**(02), 107–116 (1998)
20. Hochreiter, S., Bengio, Y., Frasconi, P., Schmidhuber, J.: Gradient flow in recurrent nets: the difficulty of learning long-term dependencies. In: A Field Guide to Dynamical Recurrent Neural Networks. IEEE Press (2001)
21. Pascanu, R., Mikolov, T., Bengio, Y.: On the difficulty of training recurrent neural networks. In: International Conference on Machine Learning, pp. 1310–1318. PMLR (2013)
22. Hochreiter, S., Schmidhuber, J.: Long short-term memory. Neural Comput. **9**(8), 1735–1780 (1997). https://doi.org/10.1162/neco.1997.9.8.1735
23. Shen, S., Van Beek, V., Iosup, A.: Statistical characterization of business-critical workloads hosted in cloud datacenters. In: 2015 15th IEEE/ACM International Symposium on Cluster, Cloud and Grid Computing, pp. 465–474. IEEE, Shenzhen, May 2015. https://doi.org/10.1109/CCGrid.2015.60

Author Index

Printed in the United States
by Baker & Taylor Publisher Services